Research and Development Activity in U.S. Manufacturing

Research and Development Activity in U.S. Manufacturing

Albert N. Link

PRAEGER

PRAEGER SPECIAL STUDIES • PRAEGER SCIENTIFIC

Library of Congress Cataloging in Publication Data

Link, Albert N
   Research and development activity in U. S. manufacturing.

   Bibliography: p.
   Includes index.
   1. Research, Industrial--United States. 2. United States--Manufactures. 3. Technological innovations--United States. I. Title.
HC110.R4L56   658.5'7'0973   80-21542
ISBN 0-03-057677-6

Published in 1981 by Praeger Publishers
CBS Educational and Professional Publishing
A Division of CBS, Inc.
521 Fifth Avenue, New York, New York 10175 U.S.A.

© 1981 by Praeger Publishers

All rights reserved

0123456789 145 987654321

Printed in the United States of America

for Carol and Jamie

ACKNOWLEDGMENTS

This study has benefited from the comments and suggestions of a number of individuals. First, I am pleased to acknowledge the helpful guidance of my colleagues, Richard Higgins and Frank Scott, through the various stages of this research. I am also grateful for the suggestions made by Rolf Piekarz, Gene Stanaland, Gregory Tassey, and Eleanor Thomas. Finally, I appreciate the comments, made by participants at the National Bureau of Standards Economics Workshop, in regard to portions of this work. Of course, I am responsible for any shortcomings.

Special thanks are due to Beth Nunn for her assistance in gathering the primary data used herein, and to Bess Yellen and Gail Micsinszki for their conscientious efforts in preparing preliminary drafts of the text.

CONTENTS

ACKNOWLEDGMENTS		vii
LIST OF TABLES		xi
LIST OF FIGURES		xiii
INTRODUCTION		xv
1	AN OVERVIEW OF THE NATURE OF R&D	1
	Innovations and Inventions	1
	Research and Development	3
	National Trends in R&D	4
	Industrial Trends in R&D	7
	Conclusions	10
2	DESCRIPTION OF THE DATA SET	11
	The Survey Questionnaire	12
	The Sample of Respondents	13
	General Description of the Data	17
3	FIRM SIZE AND INNOVATIVE ACTIVITY	25
	A Review of Previous Studies	26
	The Analytical Framework	31
	Conclusions	49

4	RATES OF RETURN TO R&D EXPENDITURES	51

    A Review of Previous Studies   52
    The Analytical Framework   55
        The Empirical Model   55
        The Statistical Estimation of
            Rates of Return   58
    A Reformulation of the Schumpeter Hypothesis   64
        Structural Change in the Rate-of-Return
            Equation   65
        Estimation of Rates of Return in
            Alternative-Size Firms   69
    Conclusions   73

5	FEDERAL SUPPORT OF INDUSTRIAL R&D	75

    A Rationale for Federally Supported R&D   78
    The Social Gains from Federally Sponsored R&D   80
    Alternative Estimates of the Return from
        Federal R&D Projects   83
    Conclusions   88

6	SUMMARY	91

APPENDIX
    A   Survey Questions   93
    B   Listing of Firms in the Sample by
            Four-Digit SIC Codes   97
    C   Logarithmic Regression Results   105

BIBLIOGRAPHY   113

INDEX   121

ABOUT THE AUTHOR   125

LIST OF TABLES   Page

1.1 Distribution of Total Industrial R&D Expenditures
     by Character of Work: 1956-77   9

2.1 Size Distribution of the Sample of 174 Firms   13

2.2 Industry Distribution and Coverage Ratios
     for Sample   14

2.3 Distribution of All Manufacturing R&D Budgets
     and Sample R&D Budgets by Industry   16

2.4 Mean Percentages of Total R&D Expenditures
     Originating from Company Funds, by Industry   18

2.5 Ten Leading Companies Receiving Federal R&D
     Contracts, Fiscal Year 1977   19

2.6 Allocation of Company-Funded R&D Expenditures
     by Character of Use   20

2.7 Average Time Lag Associated with R&D Projects   22

3.1 Regression Results from Equations 3.1-3.4:
     Sample of 174 Firms   33

3.2 Division of Sample Firms into Industry Categories   36

3.3 Regression Results from Equations 3.1-3.4, with
     11 Separate Industry Effects: Sample of 174 Firms   39

4.1 Estimated Rates of Return to R&D Expenditures by
     Character of Use: Sample of 174 Firms   59

4.2 Estimated Rates of Return to R&D Expenditures by
     Character of Use and by Industry Groupings:
     Sample of 174 Firms   60

4.3 Estimated Rates of Return to R&D Expenditures by
     Character of Use: Sample of 33 Chemical Firms   62

4.4	Estimated Rates of Return to R&D Expenditures by Character of Use: Sample of 34 Machinery Firms	63
4.5	Estimated Rates of Return to R&D Expenditures by Character of Use: Sample of 19 Transportation-Equipment Firms	63
4.6	Size Levels for Dividing Each Sample into Behavioral Regimes	69
4.7	Estimated Rates of Return to R&D Expenditures by Character of Use and by Size of Firm: Sample of 33 Chemical Firms	70
4.8	Estimated Rates of Return to R&D Expenditures by Character of Use and by Size of Firm: Sample of 34 Machinery Firms	71
4.9	Estimated Rates of Return to R&D Expenditures by Character of Use and by Size of Firm: Sample of 19 Transportation-Equipment Firms	72
5.1	Percent of Total Federal Budget Allocated to R&D Activities	76
5.2	Federal R&D Budget by Performer and by Character of Use	77
5.3	Estimated Rates of Return: Sample of 33 Chemical Firms	85
5.4	Estimated Rates of Return: Sample of 34 Machinery Firms	86
5.5	Estimated Rates of Return: Sample of 19 Transportation-Equipment Firms	87
C.1	Regression Results from Logarithmic Specifications of Equations 3.1-3.4: Sample of 174 Firms	107
C.2	Regression Results from Logarithmic Specifications of Equations 3.1-3.4, with 11 Separate Industry Effects: Sample of 174 Firms	108

LIST OF FIGURES

		Page
1.1	National R&D Expenditures: 1956-77	5
1.2	National R&D Expenditures by Character of Work: 1956-77	6
1.3	Industrial R&D Expenditures: 1956-77	8
3.1	Graphical Representation of Equations in Table 3.1: Total R&D	34
3.2	Graphical Representation of Equations in Table 3.1: Basic	34
3.3	Graphical Representation of Equations in Table 3.1: Applied	35
3.4	Graphical Representation of Equations in Table 3.1: Development	35
3.5	Graphical Representation of Chemical-Industry Equations: Total R&D	42
3.6	Graphical Representation of Chemical-Industry Equations: Basic	42
3.7	Graphical Representation of Chemical-Industry Equations: Applied	43
3.8	Graphical Representation of Chemical-Industry Equations: Development	43
3.9	Graphical Representation of Machinery-Industry Equations: Total R&D	44
3.10	Graphical Representation of Machinery-Industry Equations: Basic	44
3.11	Graphical Representation of Machinery-Industry Equations: Applied	45
3.12	Graphical Representation of Machinery-Industry Equations: Development	45

3.13	Graphical Representation of Transportation-Equipment-Industry Equations: Total R&D	46
3.14	Graphical Representation of Transportation-Equipment-Industry Equations: Basic	46
3.15	Graphical Representation of Transportation-Equipment-Industry Equations: Applied	47
3.16	Graphical Representation of Transportation-Equipment-Industry Equations: Development	47
4.1	Forward and Backward Plots of Recursive Residuals: Chemical, Machinery, Transportation-Equipment Industries	66
5.1	Optimal Level of Company-Financed R&D	84

INTRODUCTION

There have been numerous studies related to the determinants of research and development expenditures. Each investigator has characterized his study by examining some unique variables hypothesized to influence the firm's or industry's decision to invest in R&D. This commonality of the research purpose is not a criticism; it serves to emphasize the importance of identifying the technological environment that is most conducive to R&D endeavors. However, each investigator errs in viewing R&D as a single homogeneous investment activity. Most are well aware that such an assumption is invalid. The apology is always issued in terms of inadequate data on the specific uses of R&D.

The analyses presented in this book represent the first empirical study of R&D activity disaggregated by its character of use. Specifically, various aspects of the use of R&D are examined by using a sample of 174 firms from the U.S. manufacturing sector. The key variable studied is firm size. The role of firm size per se is considered as a determinant of the firm's decision to invest in alternative uses of R&D; of the rate of return earned on these investments; and of the effectiveness of the federal government in increasing the efficiency of firm-specific R&D activities. As well, a review of the economic literature related to each of these topics is presented.

An overview of the nature of R&D spending, both in the economy as a whole and in the industrial sector, is discussed in Chapter 1. There, a paradigm is developed for studying R&D as an input into the process of technological advancement. Chapter 2 describes the primary data used in this study. Chapters 3, 4, and 5 report three separate investigations into the nature of R&D. In Chapter 3, the role of firm size as a determinant of R&D spending, by character of use, is considered. In Chapter 4, the rate of return earned on each category of expenditure is estimated. In Chapter 5, the impact of federally funded R&D on the firm's self-financed R&D programs is studied. Finally, Chapter 6 offers some summary remarks.

Research and Development Activity in U.S. Manufacturing

# 1 AN OVERVIEW OF THE NATURE OF R&D

Technological change is considered by many the most important factor influencing the economic growth of our capitalistic society. It is also considered a concept that is difficult to define. The term "technological change" is often associated with inventions, innovations, product improvements, and the like. Edwin Mansfield has suggested a broad definition of technology as "society's pool of knowledge regarding the industrial arts" (1968a, p. 10). If technological change is simply applied knowledge, then it is understandable why such a nonquantifiable concept has been neglected in neoclassical economics. In the 1950s the importance of technological change gained professional attention as a result of the research by Jacob Schmookler (1952), Solomon Fabricant (1954), Moses Abramowitz (1956), and Robert Solow (1957). Their findings supported the conclusion that the growth in output per man-hour in the economy was due predominantly to the application of new production technologies. Perhaps the most influential of these investigations was Solow's: he estimated, using a production-function framework, that only 13 percent of the productivity increases in the economy from 1909 to 1949 were attributable to increases in factors of production (capital and labor), and thus suggested that the unexplained 87 percent was due to technological change. Solow's work created an interest in understanding the origins of this technology, that is, in investigating the nature of the activities included in residually measured productivity growth.

INNOVATIONS AND INVENTIONS

Where does technology originate? The early neoclassical economists viewed technology as an exogenous phenomenon—some thing or event, like manna from heaven. It was Joseph Schumpeter (1939) who first noted explicitly that technology is not an exogenous phenomenon; rather, technology is the result of innovation, and innovation emanates as an endogenous activity, from the entrepreneur. He noted (1939, pp. 87-88):

> We will now define innovation more rigorously by means of the production function. . . . This function describes the way in which quantity of products varies if quantity of factors var[ies]. If instead of quantity of factors, we vary the form of the function, we have an innovation.

Quantitatively, Schumpeter was implicitly describing innovation in terms of a shift parameter in a production function:

$$Y = A F(X), \qquad (1.1)$$

where output (Y) is some function of inputs (X), and where A, which is a function of the rate of innovation, is a coefficient that alters the form or position of the function. It is interesting to note that the concept of innovation is as broadly defined even today. For example, Howard Nason (1979, p. 69) refers to innovation as "the total process of creating, developing, and bringing to the market a new product or process." Similarly, Karl Stroetmann (1979, p. 93) contends that the term "technological innovation" refers to "the processes that go from conception or generation of an idea to its wide-scale utilization by society, including activities involved in the creation, research, development, and diffusion of new and improved products, processes, and services for private and public use." Still, none of these definitions addresses the ultimate sources of innovation.

One of the most complete inquiries into the theoretical relationship between technology, innovation, and the sources of innovation was by Abbott Usher (1954). He wrote that technology is the result of innovation, and innovation the result of invention. An invention results as the emergence of "new things" that require an "act of insight" beyond the normal exercise of technical or professional skills. Usher defined this act of insight in terms of both unlearned activities that result in a new organization of prior knowledge, and experiences coming from learned activities. Hence, if the role of the Schumpeterian entrepreneur is to innovate, he must first act as a catalyst to spur invention.

The seminal research of John Jewkes, David Sawers, and Richard Stillerman (1969, p. 169) illustrates that many of the major inventions in the past century have come from private inventors, often by accident:

> There is nothing in the history of technology. . . .
> to suggest that infallible methods of invention have
> been discovered or are, in fact, discoverable; . . .
> chance still remains an important factor in inven-
> tion and the intuition, will and obstinacy of individ-
> uals spurred on by the desire for knowledge, re-
> nown or personal gain [remain] the greatest driving
> forces in technical progress.

Since the 1920s, there has been a strong trend toward institutionalizing the process of invention. Firms, in pursuit of profit, have established industrial-research departments, instead of relying on the activities of private inventors. There are advantages in economies of scale in research, which can reduce the costs associated with the search for technical or scientific advancement. Hence, industrial firms have predominated in invention and innovation throughout modern industrial history. It is this industrial activity that may represent the primary input into the process of technological change.

Although the institutionalization of inventive activity has increased the availability of resources for scientists and inventors, it has not eliminated the aspect of risk and uncertainty from the innovation process. According to Frank Knight (1921), there is a distinction between risk and uncertainty. Risk depends on the homogeneity of events and the subsequent calculation of probabilities of occurrence; uncertainty, on the other hand, is a factor that cannot be described by objective probabilities (Kay, 1979). Although there is risk associated with most types of investment behavior, innovation is characterized by its uncertainties. Christopher Freeman (1974) lists at least three of these uncertainties: business, technical, and market. Business uncertainties relate to the unpredictable future of an innovation—specifically, how an innovation may be affected by future political, legal, and economic events. Technical and market uncertainties are project specific, and refer to the successful outcome of research expenditures and the commercial successfulness of applied innovations (timing), respectively (Kay 1979).

RESEARCH AND DEVELOPMENT

Industrial research and development (R&D) expenditures are a primary input into the innovation process. Research and development encompasses a myriad of activities. Generally, research is defined as the primary search for technical or scientific advancements; development is the translation of these advancements into product or process innovations. More specifically, the National Science Found-

4 / AN OVERVIEW OF THE NATURE OF R&D

ation (1976b) defines R&D to include activities performed by persons trained (either formally or by experience) in physical sciences, including related engineering areas, and in biological sciences, including medicine (but not psychology), if the purpose is to do one or more of the following:

1. pursue a planned search for new knowledge, whether or not the search has reference to a specific application;
2. apply existing knowledge to the creation or evaluation of a new product or process;
3. apply existing knowledge to improvement of existing products or processes.

R&D activities are formally divided by type into basic research, applied research, and development. Basic research represents original investigation, for the advancement of scientific knowledge, that does not have a specific commercial objective. Applied research, on the other hand, represents investigation directed to finding new scientific knowledge that has a specific commercial objective—product related or process related. Development is that technical activity concerned with nonroutine problems encountered in translating research findings and general scientific information into products or processes (National Science Foundation 1979b).

Clearly, R&D is not a homogeneous category. It refers to at least three easily identifiable forms of innovative activities. It is, in practice, even more varied than that. The fields associated with basic research range from the atmospheric sciences to the geological sciences. Likewise, applied-research and development projects relate to product groups as diverse as the activities of the industrial sector—the product groups range from atomic-energy devices to computing and accounting machines, from construction equipment to bicycles and parts. However, R&D is usually viewed, in both theoretical and empirical research, as a single undefined (or maybe undefinable) homogeneous investment activity. This may simply reflect our ignorance of the overall innovation process.

National Trends in R&D

As shown in Figure 1.1, national R&D expenditures increased in current dollars over the entire period from 1956 to 1977. In constant (1972) dollars, however, the increase peaked around 1969 and only recently returned to that level. A principal characteristic of R&D is that it has two primary sources of funding—the federal government and the private sector. From 1956 to 1977, federal expenditures were

FIGURE 1.1

National R&D Expenditures: 1956–77

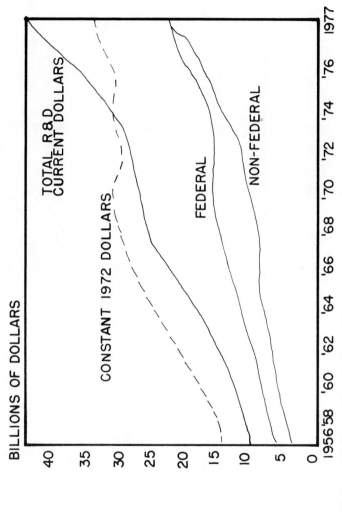

Source: National Science Foundation 1978.

FIGURE 1.2

National R&D Expenditures by Character of Work: 1956-77

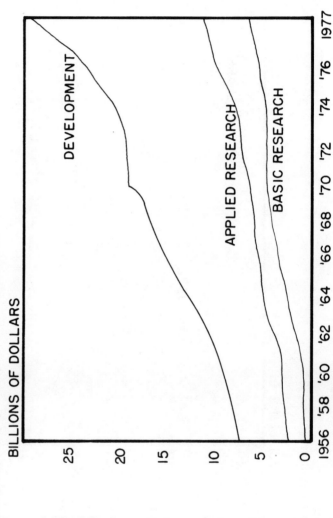

Source: National Science Foundation 1978.

greater than private expenditures in both current and constant dollars. The greatest gap between these came as a result of the escalated military spending during the mid-1960s. The dollar contribution from each sector was about the same in 1977.

As discussed above, it is more meaningful to refer to R&D by its character of use. In Figure 1.2, the spending patterns on basic, on applied, and on development are shown. In 1977, over 65 percent of total R&D expenditures went to development activities. Generally, over this period, the average annual rate of growth in development expenditures was less than that in either basic or applied. For example, from 1961 to 1967, basic expenditures grew in nominal terms at an annual rate of 13.8 percent; applied, at 7.8 percent; and development, at 7.6 percent.

Industrial Trends in R&D

The major nonfederal performer in R&D is industry, which accounted for approximately 70 percent of the total expenditures in 1977. As seen from Figure 1.3, industrial R&D increased in nominal terms over the 1956-77 period, reaching a maximum in real terms around 1969. The share of industrial R&D that was federally funded increased to a high of 59 percent in 1959 and remained constant to 1963. After 1963, that share fell steadily. In 1977, company-funded R&D accounted for about 65 percent of the total R&D performed by industry.

The distribution of industrial expenditures by character of work is illustrated in Table 1.1. The distribution of total industrial R&D expenditures in 1977 was: 78.1 percent allocated to development, 18.9 percent to applied research, and 3.0 percent to basic research. These shares have remained relatively constant over the period shown, with perhaps a decrease in basic research in favor of development. In 1962, 4.2 percent of R&D was allocated to basic research and 74.4 percent to development.

The decrease in the relative share of basic research may be associated with the growing concern that productivity and innovation are declining throughout the economy in general and in the industrial sector in particular (<u>Business Week</u> 1976, 1978; Young 1979). Is the United States losing its technological superiority? The environment for innovations has changed over the last decade. Industrial firms are decreasing their R&D efforts and diverting existing resources toward short-term activities aimed at product improvement and product imitation (defensive R&D), not process augmentation or new product development. Increased government regulation, rising capital costs,

FIGURE 1.3

Industrial R&D Expenditures: 1956-77

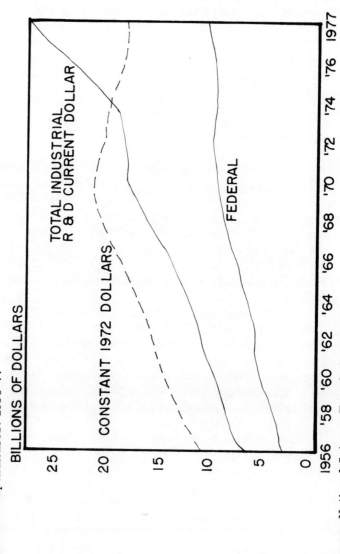

Source: National Science Foundation 1979b.

TABLE 1.1

Distribution of Total Industrial R&D Expenditures by Character of Work: 1956-77

Year	Percent Basic Research	Percent Applied Research	Percent Development
1956	3.8	19.2	77.0
1957	3.5	21.6	74.9
1958	3.5	22.8	73.7
1959	3.3	20.7	76.0
1960	3.6	19.3	77.1
1961	3.6	18.2	78.2
1962	4.2	21.4	74.4
1963	4.1	19.5	76.4
1964	4.1	19.2	76.7
1965	4.2	18.7	77.1
1966	4.0	18.3	77.7
1967	3.8	17.8	78.4
1968	3.7	17.9	78.4
1969	3.4	18.0	78.6
1970	3.3	19.0	77.7
1971	3.2	18.6	78.2
1972	3.0	18.0	79.0
1973	3.0	18.0	79.0
1974	3.1	18.7	78.2
1975	3.0	18.9	78.1
1976	3.1	18.9	78.0
1977	3.0	18.9	78.1

Source: National Science Foundation 1979b.

and general market uncertainties have made basic innovative programs—those primarily responsible for technological growth—unprofitable.

CONCLUSIONS

If one adheres to the paradigm that technological change is a result of innovation and invention, and that innovation and invention are one output resulting from industrial R&D, then it seems that an understanding of the nature of R&D is fundamental for a theoretical or practical understanding of economic growth in the United States. An approach toward that understanding should view R&D not as a composite category, but rather, by its character of use. That is the goal of the following chapters.

2 DESCRIPTION OF THE DATA SET

This chapter discusses the primary data used in this study. The data are from a sample of 174 manufacturing firms for which information on basic-research, applied-research, and development expenditures is available. From a descriptive analysis of these data, it is concluded that the sample is representative of R&D activity within the manufacturing sector.

There are few published data sources reporting R&D statistics at either the industry or the firm level. The best-known and most accessible source is the National Science Foundation's annual publication, Research and Development in Industry. R&D statistics corresponding to selected two-digit- and three-digit-level SIC manufacturing industries—total and by character of use—are available. This publication provides a complete source of data on industrial R&D activity. The Compustat data files of Investor's Management Sciences, Inc., a subsidiary of the Standard and Poor's Corporation, reports annual R&D expenditures for all major public companies. This latter source has been used extensively and is attractive to researchers since Compustat also reports the corresponding financial information from each firm's balance sheet and income statement. As well, Business Week publishes these Compustat estimates for the largest industrial firms in its annual R&D Scoreboard.

The National Science Foundation's data are not appropriate for this study, nor are the Compustat files, since firm data disaggregated by character of use of R&D are required. Therefore, a sample of firms was selected and surveyed. The initial sample consisted of all industrial firms in the manufacturing sector listed in the 1977 Fortune 1,000. It was anticipated that many of these firms would not have R&D programs; however, to prevent biasing the survey toward the very largest firms, by an a priori judgment of what size range

12 / DESCRIPTION OF THE DATA SET

of firms to consider, the initial sample was kept near 1,000. The chief executive officer in each firm was sent four items: a cover letter describing the type of information requested and ensuring the confidentiality of responses; the survey instrument, accompanied by a list of National Science Foundation definitions of related terminology; and a self-addressed, stamped envelope.

THE SURVEY QUESTIONNAIRE

Each of the four questions applicable to this study are reproduced in Appendix A. The questions were formulated to accommodate answers in percentage terms; this was done primarily to simplify completion of the questionnaire. Since much of the information requested might be of a proprietary nature, it was anticipated that firms would be more willing to respond if percentages, rather than actual dollar figures, were requested. All requested data refer to the 1977/78 fiscal year.

A total of 359 firms responded to the survey (and the followup); however, only 272 were willing to participate, and thus returned a completed or partially completed questionnaire. In general, the 87 firms that responded, but were not willing to participate, reported either that the information requested was proprietary in nature and would not be released (52 firms), or that company policy precluded participation (35 firms). Presumably, these reasons also apply to firms that did not return the questionnaire; or those firms may simply not be involved in R&D. Of the returned questionnaires, 98 were not completed (generally, questions 2 and 3 were not answered), and thus could not be used here. The final sample consisted of 174 manufacturing firms.

The primary problem with this survey instrument is the appropriateness of the categories labeled basic research, applied research, and development. Since the early 1950s, the National Science Foundation has used these categories in an attempt to compare R&D expenditures in all sectors of the economy with a common set of definitions. Distortions have occurred with this procedure. Howard Nason, Joseph Steger, and George Manners (1978) addressed this question with respect to basic research, and concluded that the category labeled basic research is not appropriate for describing industrial R&D activity. They contend that firms do not think in such terms—firms categorize R&D activity to meet their own organizational purposes. Nevertheless, most organizations use the National Science Foundation definitions for external purposes, and thus, according to Nason, Steger, and Manners, the category basic research does have "construct validity" (as do applied research and development); and the definition is applicable and usable in surveys for making interfirm comparisons.

THE SAMPLE OF RESPONDENTS

The firm names and four-digit SIC codes for the sample of 174 firms are reported in Appendix B. As a sample, these firms account for 36.9 percent of all company-financed R&D in the manufacturing sector. Of the three manufacturing firms leading in R&D expenditures (General Motors, Ford Motor Company, and International Business Machines), only General Motors is in this sample. As well, two of the largest ten firms, and 11 of the largest 25 firms, conducting R&D programs are represented.

Since size will be examined in the later chapters as an important firm characteristic related to R&D activity, it is important that the size distribution of the sample is representative of the entire manufacturing sector. The percentages of the sample in various strata of the Fortune 1,000 list are shown in Table 2.1. It appears that a greater proportion of the relatively larger-sized firms responded to the sample questionnaire. This response distribution acknowledges the fact that R&D spending is positively correlated with firm size (sales) per se and that the smallest firms in the Fortune 1,000 are not classified within the manufacturing sector. Still, the firms in the sample accounted for 42.2 percent of total net sales in manufacturing.

TABLE 2.1

Size Distribution of the Sample 174 Firms

Distribution of Firms in Fortune 1,000 List	Percentage of Sample of 174 in Each Stratum
Largest 50 firms	13.8
Next 100 firms	21.3
Next 150 firms	13.8
Next 200 firms	13.2
Next 500 firms	37.9

TABLE 2.2

Industry Distribution and Coverage Ratios for Sample

Industry	SIC code	Firms in Sample	Net Sales Coverage Ratio	R&D Coverage Ratio
Food and kindred products	20	16	33.2%	55.2%
Textiles and apparel	22, 23	10	35.0	58.4
Lumber, wood products, and furniture	24, 25	2	22.3	30.5
Paper products	26	5	10.0	20.3
Chemicals	28	33	55.1	40.5
Petroleum products	29	11	74.5	54.9
Rubber products	30	7	26.7	39.2
Clay, glass, and stone	32	7	42.6	64.3

Primary metals	33	5	25.7	26.0
Fabricated metals	34	2	1.5	0.7
Machinery	35	34	49.6	36.1
Electrical equipment	36	12	26.4	18.6
Transportation equipment	37	19	56.8	54.2
Instruments	38	9	28.0	28.6
All others	21, 27, 31, 39	2	1.8	0.4
All industry		174	42.2	36.9

TABLE 2.3

Distribution of All Manufacturing R&D Budgets
and Sample R&D Budgets by Industry

Industry (SIC code)	Percent of All Manufacturing R&D Budgets	Percent of Sample R&D Budgets
20	2.1	2.7
22, 23	0.8	0.7
24, 25	0.9	0.5
26	1.7	0.6
28	16.3	17.9
29	4.3	6.3
30	2.0	1.9
32	1.5	2.3
33	2.8	2.0
34	1.9	0.1
35	17.5	17.1
36	18.3	9.2
37	22.9	33.5
38	6.9	5.2
21, 27, 31, 39	0.1	0.0
	100.0	100.0

R&D ACTIVITY IN U.S. MANUFACTURING / 17

The distribution of the 174 firms among 15 industry groupings is reported in Table 2.2. The chemical, machinery, and transportation-equipment industries are represented by the largest number of firms. To get a better idea of the representativeness of each industry grouping, coverage ratios were calculated for both net sales and company-financed R&D expenditures. Data for sales and R&D expenditures are available from Compustat, by firm, for the 1976/77 fiscal year. The industry totals were taken from <u>Research and Development in Industry</u>. In several instances, industry R&D figures were estimated when not reported.

In general, the sample of 174 firms is extremely representative of R&D performance in manufacturing. The exceptions are in industry groupings containing a small number of firms. The five leading R&D industries in the manufacturing sector are chemicals, petroleum products, machinery, electrical equipment, and transportation equipment. The R&D coverage ratio for these five industries as a group is 39.1 percent.

The interindustry distribution of R&D expenditures for the sample also compares favorably with that of the entire manufacturing sector. For example, the chemical industry accounts for 16.3 percent of all manufacturing R&D expenditures and 17.9 percent of expenditures for the sample. Likewise, the machinery industry accounts for 17.5 percent of all manufacturing R&D expenditures and 17.1 percent of sample R&D expenditures. Similar calculations for each of the 15 industry groupings are reported in Table 2.3.

GENERAL DESCRIPTION OF THE DATA

The 174 firms' responses to question 1 are reported in Table 2.4. The mean percentage of total firm R&D that is company financed is, for the sample of 174 firms, 93.1 percent. This figure is high compared with the manufacturing sector as a whole, where 65 percent of total R&D was company financed in 1977. The distortion is due to the absence of those firms whose R&D programs are financed mostly through federal R&D contracts. The anomaly occurs primarily in the electrical-equipment industry (SIC 36) and in the transportation-equipment industry (SIC 37). As seen from Table 2.5, many of the leading companies receiving federal R&D support from either the Department of Defense or the National Aeronautics and Space Administration are in the sample. However, these firms rely relatively less on federal support for R&D than do firms receiving a smaller part of total federal allocations. These R&D

TABLE 2.4

Mean Percentages of Total R&D Expenditures
Originating from Company Funds, by Industry

Industry (SIC Code)	Mean Percent of Company-Funded R & D
20	99.9
22, 23	99.8
24, 25	100.0
26	100.0
28	98.2
29	95.1
30	86.9
32	98.9
33	96.4
34	100.0
35	94.2
36	83.0
37	69.8
38	99.7
21, 27, 31, 39	100.0
All industry	93.1

performers with a smaller part of federal aid are somewhat under-represented in the sample.

The allocation of company-funded R&D expenditures by character of use, as derived from survey question 2, is reported in Table 2.6. The sample means are: 3.9 percent allocated to basic research, 29.6 percent to applied research, and 66.5 percent to development. These estimates compare favorably with the 1977 distribution pattern reported by the National Science Foundation for the entire manufacturing sector: 3.6 percent to basic research, 22.1 percent to applied research, and 74.3 percent to development. Two interesting generalizations can be made from these statistics: every industry is least intensive in basic research;

TABLE 2.5

Ten Leading Companies Receiving Federal R&D
Contracts, Fiscal Year 1977

Company	Department of Defense Awards Percent of Total Dollars Awarded
Rockwell International*	35.64
McDonnell Douglas*	4.88
Martin Marietta	4.21
Bendix*	3.19
General Dynamics	2.77
General Electric*	2.42
Lockheed Electronics*	2.40
International Business Machines	2.33
Thiokol	2.20
Boeing*	1.87

Company	NASA Awards
McDonnell Douglas*	5.11
Lockheed*	3.32
United Technologies	3.15
Boeing*	3.14
General Electric*	3.02
Rockwell International*	2.94
Grumman	2.83
General Dynamics*	2.72
Hughes Aircraft	2.17
Northrop	2.08

*In the sample 174 firms.
Sources: U.S. Department of Defense, "100 Companies Receiving the Largest Dollar Volume of Military Prime Contract Awards, Fiscal Year 1977" (Washington, D.C., 1978), and National Aeronautics and Space Administration, "Annual Procurement Report, Fiscal Year 1977" (Washington, D.C., 1978).

TABLE 2.6

Allocation of Company-Funded R&D
Expenditures by Character of Use

Industry (SIC code)	Mean Percent Basic	Mean Percent Applied	Mean Percent Development
20	9.2	36.3	56.5
22, 23	3.6	13.8	82.6
24, 25	0.0	60.0	40.0
26	4.0	38.0	58.0
28	5.6	38.8	55.6
29	4.4	39.4	56.2
30	6.3	29.8	64.9
32	2.7	38.0	59.3
33	10.8	29.0	60.2
34	2.5	37.5	60.0
35	3.2	21.8	75.0
36	1.2	20.8	78.0
37	2.4	26.8	70.8
38	1.4	19.3	79.3
21, 27, 31, 39	16.5	27.0	56.5
All industry	3.9	29.6	66.5

and every industry, except for lumber, wood products, and furniture (SICs 24-25), is most intensive in development activity.

Compustat reports total company-financed R&D expenditures for each of the 174 sample firms for the 1976/77 fiscal year (the latest data available at the time this study was begun). The firm percentages for basic research, applied research, and development were multiplied by each firm's R&D budget to obtain the dollar amount allocated by each firm to these three categories. These dollar amounts represent the primary data for this study.

The responses from questions 3 and 4 were not used in the analyses in Chapters 3, 4, and 5; however, they are discussed here briefly primarily because of their uniqueness as a firm-specific data set.

Of the 174 firms in the sample, 58 reported that they received federal support for R&D, presumably through government contracts. Although these few firms cannot be meaningfully analyzed on an industry-specific basis, two general observations can be made. First, the distribution of federally supported R&D by character of use among these 58 firms is: 4.2 percent to basic research, 16.1 percent to applied research, and 79.7 percent to development. This distribution parallels the 1977 manufacturing percentages, as reported by the National Science Foundation—2.0 percent to basic research, 13.1 percent to applied research, and 84.9 percent to development. Second, the correlation between the percentage of company funds allocated to basic research and the percentage of federal funds allocated to basic research is 0.292 (significant at the .05 level). The correlation is 0.244 (significant at the .05 level) for applied research and 0.355 (significant at the .01 level) for development.

As explained in the survey questionnaire, the time lag referred to in question 4 corresponds to the period between the firm's initial R&D expenditures and the implementation or commercialization of the result of the project, in the form of an improved product or process. The average lags reported in Table 2.7 meaningfully relate to applied-research and development expenditures. The mean average lag associated with R&D for the entire sample is 3.9 years. There is little existing literature that provides for a comparison of these estimates. John Enos (1962) studied more than 30 inventions from the late 1800s and early 1900s, and estimated that the lag between invention and innovation is about 11 years in the petroleum and about 14 years in other durable-goods industries, with standard deviations of five years and 16 years, respectively. But the nature of the innovative process is different today. R&D projects are more short run in nature, with specific product- and/or process-related goals. Therefore, it is not unreasonable to expect that the lag has been

TABLE 2.7

Average Time Lag Associated
with R&D Projects

Industry (SIC codes)	Mean Average Lag (years)
20	3.1
22, 23	3.3
24, 25	3.5
26	5.2
28	4.2
29	5.6
30	3.6
32	4.1
33	4.2
34	2.8
35	3.8
36	3.0
37	4.0
38	4.2
21, 27, 31, 39	5.0
All industry	3.9

reduced substantially. Leonore Wagner (1968) examined survey data on 36 manufacturing firms, and found that the mean length of project duration for successful applied-research and development projects was around two years in nondurables industries and three years in durables industries. These findings are similar to those reported in Table 2.7.

It is interesting to note that there is industry variation in these reported lags. The standard deviation is 2.1 years. The longest lags are reported for the petroleum (SIC 29) and paper (SIC 26) industries, and the shortest, for the food (SIC 20) and the textile and apparel (SICs 22-23) industries. The correlation between the percentage of each firm's R&D budget allocated to development and the average lag is -0.230 (significant at the .01 level). The corresponding correlation between the average lag and the percentage allocated to applied research is 0.145 (significant at the .05 level). These findings may be interpreted as reinforcing the feeling that development activity is designed to produce results in the short run, relative to applied-research activity. Also, the correlation between the average lag and the percentage of each firm's R&D budget that is company financed (question 1) is -0.268 (significant at the .01 level), perhaps suggesting that the contribution of federal R&D contracts does not generate the type of technical efficiencies associated with short-term R&D projects. These conclusions are tentative, and the data are presented simply for descriptive purposes.

3 FIRM SIZE AND INNOVATIVE ACTIVITY

Ever since Schumpeter wrote <u>Capitalism, Socialism, and Democracy</u> (1947), many economists have been willing to associate economic growth through innovation with monopoly power and large firm size. The so-called Schumpeterian hypothesis is posited on the view that in a capitalistic system, economic growth occurs through a process of "creative destruction," whereby the old industrial structure—its product, its process, or its organization—is continually changed by new innovative industrial activity. Schumpeter (1947, p. 83) noted that such "industrial mutation . . . that incessantly revolutionizes the economic structure from within . . . is the essential fact [of] capitalism." The motivating force behind the process of creative destruction is the promise of economic profit achieved through innovation. According to Schumpeter, large firm size is essential to the success of such a process. Larger firms can provide economies of scale in production and innovation, thus making available sufficient physical and financial resources necessary for the processes' successful completion.

Schumpeter claimed the monopolistic firm will have a greater demand for innovative activity than will the competitive firm since the monopolist can profit from the innovation, as a result of its market power. Kenneth Arrow (1962) first demonstrated that this demand argument is incorrect. Arrow's analysis has been criticized by Harold Demsetz (1969), but the more recent work of Sheng Cheng Hu (1973) and Yew-Kwang Ng (1971) seems to resubstantiate it. Most empirical analyses of Schumpeter's hypothesis focus only on the supply of innovations.

In addition to bigness per se being that characteristic that approximates the firm's ability to provide sufficient resources necessary for the successful completion of R&D projects of some given size, large firms can also provide economies of scale within the R&D facility itself. The larger firms can support more effectively cooperative R&D among the many divisions of the firm, as well as attract extramural resources for use within the facility (Kamien and Schwartz 1975). Both scale phenomena are implicit in Schumpeter's argument.

During the past two decades, statistical tests of the Schumpeterian hypothesis have focused on two relationships: between firm size and innovative activity, and between market concentration and innovative activity. In this chapter, we are specifically concerned with the relationship between firm size and innovative activity. In particular, we address the general question: Is the intensity of innovative activity relatively greater in larger-sized firms?

Innovative activity will be measured in two ways: first, innovative activity will be measured as total R&D expenditures by the firm; and second, these expenditures will be disaggregated, by character of use, and measured specifically as basic-research expenditures, applied-research expenditures, and development expenditures. Accordingly, the empirical section of this chapter will investigate the relationship between these R&D expenditures, by category and firm size.

Under certain assumptions, the results of this inquiry may lead to definite policy conclusions regarding the structure of manufacturing industries. If intensity in R&D spending is a determinant of innovative success, and if innovation is in fact a significant input into technological growth—private and social—then empirical evidence to support the hypothesis that firm size is a prerequisite for R&D intensity may suggest that an industry composed of large firms (presumably with some significant market power) may be the most technologically efficient form of market structure (Kamien and Schwartz 1975).

There have been numerous tests of Schumpeter's hypothesis; however, each of these studies suffers from two general shortcomings. First, R&D activity is viewed as a single homogeneous category of behavior. As discussed in Chapter 1, this assumption is obviously inaccurate. Second, all of these earlier studies employed data at least 15 years out of date. As also noted in Chapter 1, the structure of R&D activity in industry has changed over the past decade and thus deserves a reinvestigation.

A REVIEW OF PREVIOUS STUDIES

The principal methodological problem associated with any test of the Schumpeterian hypothesis relates to the investigator's choice of an empirical measure of innovative activity. Generally, the concept of innovative activity is approximated by some measure of the firm's inputs into the innovation process. There are several inputs that can be easily identified: R&D expenditures, R&D personnel, and R&D technical scientists. None of these measures is void of problems.

R&D expenditures, for example, represent a broad category of spending that may not fully capture the firm's financial commitments to innovation. First, innovative improvements can result from activities within the firm that are not directly funded under the heading of R&D. Second, heterogeneous accounting procedures and treatments (or interpretations) of tax laws make interfirm (and especially interindustry) comparisons of R&D spending somewhat inaccurate. Similarly, the category called R&D employees is a heterogeneous grouping. One firm may employ a large R&D staff simply to accommodate government-contract obligations. The contracted output may not even be related to a form of innovative activity. Nevertheless, these input measures have gained acceptance within the literature, probably since alternative disaggregate measures are not as readily available.

Some investigators (Mansfield 1964, 1969; Mueller 1966) have assumed that patents represent a reasonable index of innovative activity; but patented inventions vary markedly in quality, and the number of patents of a firm may in fact not be related to the impact of any of its particular inventions. Moreover, according to Kennedy and Thirwall (1972, p. 45),

> "since invention is one step removed from the application of new knowledge to production a relationship between R&D and invention tells us nothing about the impact of R&D on the rate of measured technical progress."

One of the earliest studies of the relationship between R&D activity and firm size was by Henry Villard (1958). Using National Science Foundation statistics, he noted that the percentage of firms involved in R&D increased with the size of the firm. In 1956, for example, 94 percent of firms with more than 5,000 employees were involved in R&D; but only 8 percent of firms with less than 100 employees were involved in R&D.

Schmookler (1952), however, questioned the meaning of Villard's results. First, he noted that Villard's use of National Science Foundation data biased his sample toward large firms with well-established R&D laboratories. Small firms may conduct R&D, but do not classify it under such a category. Second, Villard's results do not necessarily imply that increased firm size causes greater R&D activity; a reverse causation may apply as well. It is interesting to note that these types of interpretative criticisms are present in the literature for the two decades of research that followed.

Similarly, James Worley (1961) stated that the issue in question is the proper test of the relationship between firm size, bigness per se, and R&D activity. It is not clear that economists today would be as ambitious as Worley was in suggesting that there is just one proper test of this proposition. Using a sample of 198 firms, grouped into eight industries, from the 1955 Fortune 500, he tested empirically the following relationship: $RD = A \cdot (SIZE)^b$. Using a log-linear transformation, measuring RD as the firm's R&D personnel, and measuring SIZE as the firm's total employment figure, he found that the elasticity of R&D with respect to firm size (b) was statistically greater than unity only for firms in the petroleum and electrical-machinery industries.

Ira Horowitz (1962), also found weak statistical evidence of a relationship between R&D activity and firm size. The correlation between output per establishment in manufacturing industries and R&D participation, for the period 1951-52, was positive but not statistically significant. He, like Worley, concluded (1962, p. 300) that at best there may be "some empirical justification for the contention that bigness favors research, though the relationship might be working in the other direction as well."

Daniel Hamberg (1964, 1966) stated that the purpose of his study was to present empirical evidence in support of the "giant firm." His argument, or "new defense," was that "cheap and simple inventions" have already occurred, and that therefore, any technical advancements would only be possible with the physical and financial resources available in large industrial firms. His empirical support of the thesis that R&D activity is an increasing function of firm size was based on rank-order correlation coefficients; and on least-squares analyses (patterned after Worley) of the relationship between R&D personnel and firm size, and of that between R&D personnel, divided by total employment—a measure of intensity—and firm size, for 17 industry groups, based on a sample of 340 firms from the 1960 Fortune 500 list. All of his empirical evidence provides weak support for the virtues of the role of the giant firm in the area of technological advancement.

William Comanor (1967) also estimated the elasticity of R&D with respect to firm size, using the models of Worley and Hamberg for a sample of 387 firms, grouped into 21 industries, from the 1955 and the 1960 Fortune 500. The empirical results were similar to those of the earlier studies. Going one step further, Comanor (1967, p. 645) regressed these calculated industry elasticities against the average firm size for each industry and found that where "firms tend generally to be small, the estimated elasticities are low, and these coefficients, on average, are less than one. Where firms are large, however, the estimated elasticities are higher and approximately equal to one." This finding led Comanor to conclude that if increasing returns to scale do exist within an industry, they exist only to a point and then level off or perhaps even decline.

Frederic Scherer (1965b) strongly criticized, on methodological grounds, the analytical approach used by both Worley and Hamberg. He said, first, their conclusions are based on a blind adherence to significance tests at the .05 level, without an admission that the numerical results they obtained were, nevertheless, best estimates. Second, the samples chosen by Worley and Hamberg were biased in favor of the largest R&D performers, Scherer noted. Alternatively, a proper sample should contain a spectrum of variously sized firms, even those with no R&D programs. In this light, Scherer reassessed the empirical relationship between firm size and R&D personnel, using a sample of 325 firms from the 1955 Fortune 500, and estimated a cubic relationship where firm size was measured as total sales, and also as the logarithm of total sales for both the entire sample and for six separate industry groupings: food and tobacco, chemicals, petroleum products, primary metals, machinery, and electrical equipment. For five of these industries, the resulting relationships between R&D activity and firm size were nonlinear, implying that R&D intensity increases with firm size only among the smallest firms and, in some instances, medium-sized firms. However, the empirical evidence for the chemical industry does suggest that R&D intensity increases with firm size over all size levels. Scherer (1965b, p. 265) concluded that his "results do, however, support the policy conclusion drawn by [Worley and] Hamberg: that gigantic scale is far from an essential condition for rigorous industrial research and development activity, and that bigness may indeed be a stifling factor."

Dennis Mueller (1967) argued that firm behavior is a complex process, and that most firm decisions are not independent events, but simultaneous decision processes. Accordingly, he attempted to model the firm's decision-making process subject to the constraint that each decision has an equal claim to the firm's financial resources. One empirical test of his model was related to the allocation of R&D expenditures within the firm. Using data from a sample of 67 firms from a cross section of the manufacturing sector over the period 1957-60, he concluded that R&D intensity decreases with firm size, ceteris paribus.

Mansfield (1968b) also found that R&D intensity decreases with firm size. His sample consisted of nine firms in the petroleum industry, eight firms in the drug industry, seven firms in the steel industry, and four firms in the glass industry. There was, however, statistical evidence from a sample of ten chemical firms that intensity increases with firm size. As well, it appeared that each of these industry relationships was stable over the sample period, 1945-59.

Finally, Peter Loeb and Vincent Lin (1977) reestimated the relationship between R&D activity (measured both as R&D expenditures and as average annual compensation paid to R&D scientists and engineers) and firm size (measured both as sales and as income) for a composite of data from six pharmaceutical firms over the years 1961-72. Using a specification-error approach, they concluded that the best specification is a quadratic relationship between R&D expenditures and sales. Their results also supported the earlier findings that R&D intensity diminishes as firm size increases.

It should be clear from these studies that there is very little support for the hypothesis that R&D intensity increases with firm size. Other studies reaching similar conclusions are by William Adams (1970) and Louis Phlips (1971). The only exception appears to be the behavior observed among firms in the chemical industry. Most studies do, however, suggest that the relationship between R&D activity and firm size is nonlinear, and that the importance of bigness per se increases only among the smallest-sized firms.

Henry Grabowski and Mueller (1970) have suggested that this lack of empirical verification perhaps rests in the inadequacy of data for measuring innovative activity, as others have previously suggested, or in the fact that the large industrial-research complexes of today have changed the structure of capitalism that Schumpeter originally envisioned. An alternative explanation was first proffered by Jesse Markham (1965), and then more recently reinforced by Franklin Fisher and Peter Temin (1973). Fisher and Temin contend that the existing empirical studies relating innovative activity, as measured by some absolute index like R&D expenditures, to firm size relate to a different hypothesis than the one suggested by Schumpeter. The quantitative argument developed by Fisher and Temin is basically that a positive and increasing relationship between innovative inputs and firm size—the general empirical test—is neither necessary nor sufficient to imply a positive and increasing relationship between innovative output and firm size—the Fisher and Temin interpretation of Schumpeter's hypothesis—given economies of scale in the production of output and innovation.

Scherer (1973, p. 1) strongly disagrees with their conclusions:

> The authors [Fisher and Temin] do violence to both Schumpeter and common sense in formulating their mathematical model; they display considerable ignorance of the literature they criticize; and they ignore results inconsistent with their arguments in papers which they allegedly did read.

According to Scherer, the basic, although erroneous, assumption underlying the Fisher and Temin criticism is that profits from innovations are synonymous with the output of innovative activity. When this conceptual error is corrected, their basis for criticism is removed.

Still, the Schumpeterian hypothesis remains important for economic investigation. If specific categories of innovative activity do increase with firm size, then arguments against firm bigness must be tempered in light of the social contributions associated with innovation and subsequent technological growth.

THE ANALYTICAL FRAMEWORK

In this section, the relationships between total R&D, basic-research, applied-research, and development intensity, and firm size are examined, using data from the survey questionnaire described in Chapter 2. The empirical model is a cubic estimating equation relating the four alternative measures of R&D to firm size. Specifically, total R&D expenditures (RD), basic-research expenditures (B), applied-research expenditures (A), and development expenditures (D) are individually related to firm size (S), as follows:

$$RD = \alpha_0 + \alpha_1 S + \alpha_2 S^2 + \alpha_3 S^3 + \epsilon_{RD} \tag{3.1}$$

$$B = \beta_0 + \beta_1 S + \beta_2 S^2 + \beta_3 S^3 + \epsilon_B \tag{3.2}$$

$$A = \gamma_0 + \gamma_1 S + \gamma_2 S^2 + \gamma_3 S^3 + \epsilon_A \tag{3.3}$$

$$D = \delta_0 + \delta_1 S + \delta_2 S^2 + \delta_3 S^3 + \epsilon_D \tag{3.4}$$

where the ϵ's are error terms assumed independent and normally distributed.

The cubic form of these estimating equations is especially useful for an investigation of this type, since that structural form allows the relationship between RD, for example, and S to be nonlinear; that is, $\partial RD/\partial S = \alpha_1 + \alpha_2'S + \alpha_3'S^2$, from which it can be discerned whether R&D intensity is an increasing or decreasing function of firm size. As well, the form of equations 3.1 through 3.4 allows for the influence of the very largest firms being studied to have a pronounced impact on the fitted regressions. This is not undesirable since, as Scherer (1965b) noted, these largest firms account for the "lion's share" of resources devoted to R&D activity and thus provide a significant impact on resulting innovations and on subsequent technological growth.

Many other investigators have used log-linear forms (Comanor 1967; Hamberg 1964, 1966; Mansfield 1968b; Worley 1961) or semi-log-linear forms (Loeb and Lin 1977; Scherer 1965a, 1965b) for this kind of estimation in order to dampen the influence of the very largest firms (as well as to improve the fit of the estimated equation). These logarithmic transformations of equations 3.1 through 3.4 will be examined here, although they are not the primary focus of the analysis, and will be reported in Appendix C.

As noted in Chapter 2, data are available from the survey questionnaire on the percentage of each firm's total R&D expenditures allocated to basic research, to applied research, and to development for each of 174 firms. The dependent variables in equations 3.1 through 3.4 represent dollar expenditures. The dollar value for each firm's total R&D expenditures was obtained from Compustat. All Compustat data relate to the 1976/77 fiscal year, and the survey percentages correspond to the 1977/78 fiscal year. In the estimation of equations 3.1 through 3.4, innovation expenditures were measured in millions of dollars, and sales were measured in billions of dollars. The ordinary least-squares results are reported in Table 3.1. The overall specification of each of the four equations is highly significant, as evidenced by the reported F-statistics. The least-squares results can be interpreted easily from a graphical representation of each equation, as shown in Figures 3.1 through 3.4. For each category of spending, the estimated relationships are clearly nonlinear. These relationships are strongly influenced, as would be expected, by the very largest R&D performers in the sample. In particular, the sample contains General Motors, with a total R&D budget of $1.26 billion, followed by General Electric and Xerox, with R&D budgets of $412 million and $226 million, respectively.

Innovation intensity can be measured from the graphs by the slope of a ray drawn from the origin to the fitted line. The equations graphically represented in Figures 3.1 through 3.4 imply that innovation intensity, as measured in terms of total R&D spending and in terms of each category of use, increases with firm size only among the smallest and the very largest firms in the sample. Maximum innovation intensity, defined as R&D spending divided by sales, is reached at sales levels between $1.0 billion and $2.0 billion. About 73 percent of the firms in this sample have sales less than $2.0 billion. Beyond these levels, there is a tendency for innovation intensity to decrease, at least to the point where the influence of the largest firms takes over as evidenced by the upswing in each of the curves, and then increase rapidly.

These empirical results correspond to the findings of Worley (1961), Hamberg (1964, 1966), Comanor (1967), and Scherer (1965a, 1965b), who examined a composite cross section of firms from the

TABLE 3.1

Regression Results from Equations
3.1-3.4: Sample of 174 Firms

Variable	Total R&D	Basic	Applied	Development
Intercept	1.313	0.294	0.486	1.174
	(0.31)	(0.60)	(0.41)	(0.38)
S	21.181*	1.401	5.514*	14.038*
	(7.29)	(4.19)	(6.70)	(6.63)
S^2	-1.268*	-0.101*	-0.212*	-0.937
	(-4.56)	(-3.14)	(-2.69)	(-4.62)
S^3	0.029*	0.002*	0.003†	0.024*
	(6.13)	(3.82)	(2.37)	(6.80)
R^2	0.864	0.549	0.591	0.893
F-level	359.054	68.876	81.965	470.397

*Significant at the .01 level.
†Significant at the .05 level.
<u>Note:</u> t-statistics are reported in parentheses below each estimated coefficient.

FIGURE 3.1

Graphical Representation of Equations in Table 3.1: Total R&D

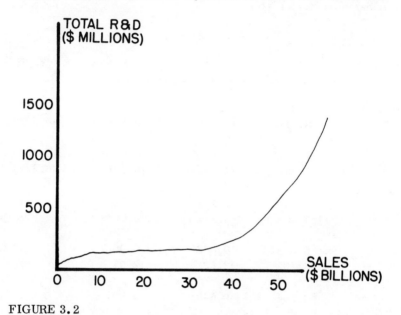

FIGURE 3.2

Graphical Representation of Equations in Table 3.1: Basic

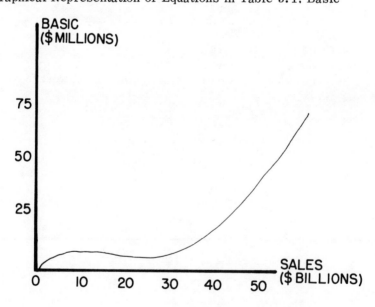

FIGURE 3.3

Graphical Representation of Equations in Table 3.1: Applied

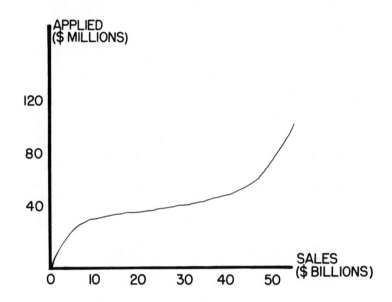

FIGURE 3.4

Graphical Representation of Equations in Table 3.1: Development

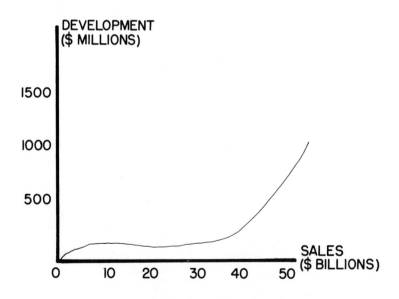

TABLE 3.2

Division of Sample Firms into Industry Categories

Industry	SIC code	Number of Firms
Food and kindred products	20	16
Tobacco and textile mill products	21, 22	11
Lumber, paper, and printing products	24, 26, 27	8
Chemicals and allied products	28	33
Petroleum and coal products	29	11
Rubber, stone, clay, and glass products	30, 32	14
Primary and fabricated metal products	33, 34	7
Machinery, except electrical	35	34
Electrical and electronic equipment	36	12
Transportation equipment	37	19
Instruments and related products	38	9
Total		174

manufacturing sector. Specifically, for each category of expenditure on innovative activity, intensity increases with firm size only among the smallest and the very largest firms. In general, firm size was not a prerequisite for R&D intensity across firms of all sizes.

It is misleading to generalize about the impact of firm size on innovation intensity from a cross-sectional analysis of firms from heterogeneous industries. It is also somewhat unjustified to compare a dollar of R&D, regardless of category of use, in one industry with a dollar of R&D in some other industry; the nature of the activities is different in scope and in expense. In addition, the influence of the very largest firms in a particular industry composed of firms that are relatively small, compared to the size of firms from other industries, is completely lost when firms are pooled across industries. Therefore, equations 3.1 through 3.4 were estimated allowing for separate industry-intercept terms and for separate industry-slope terms. For example, equation 3.1 takes the form:

$$RD = \Sigma_i \alpha_0^i + \Sigma_i \alpha_1^i S + \Sigma_i \alpha_2^i S^2 + \Sigma_i \alpha_3^i S^3 + \epsilon_{RD}, \qquad (3.5)$$

where the superscript i refers to each of the 11 separate industry groupings defined in Table 3.2. These industry divisions are broader than those used for descriptive purposes in Chapter 2. It should also be noted that these categories are not intended to represent perfectly homogeneous groupings; rather, they are intended only as a first step toward a more disaggregated analysis. The estimated regression results for the modified versions of equations 3.1 through 3.4 are reported in Table 3.3. The colinearity problem in the estimation of these four equations is severe, as evidenced by the fact that most of the estimated coefficients are not statistically significant at conventional levels (Scherer 1965b). There is no getting around this problem: when those industries with the larger number of behavior observations (chemical, machinery, and transportation equipment) were used individually in an estimating equation, the colinearity problem remained. Even so, the estimates reported in Table 3.3 are the best estimates obtainable (in a statistical sense) for the structure of the relationship under study.

Three industries are of particular interest for comparative purposes since they have been studied by other investigators and will be examined individually later: the chemicals, machinery, and transportation-equipment industries.

The regression results (reported in Table 3.3) that correspond to the chemical industry (SIC 28) are graphically represented in Figures 3.5 through 3.8. Each of the equations is nonlinear and is influenced by the largest firms in the subsample—Dow Chemical,

TABLE 3.3

Regression Results from Equations 3.1 - 3.4, with
11 Separate Industry Effects: Sample of 174 Firms

Variable	Total R&D	Basic	Applied	Development
D20	2.843 (0.22)	0.370 (0.13)	1.896 (0.37)	0.035 (0.01)
D20(S)	-2.441 (-0.08)	-0.524 (-0.08)	-4.125 (-0.34)	3.590 (0.17)
D20(S^2)	5.560 (0.35)	0.455 (0.13)	3.562 (0.58)	0.824 (0.08)
D20(S^3)	-0.844 (-0.42)	-0.062 (-0.14)	-0.479 (-0.61)	-0.224 (-0.16)
D21, 22	-1.228 (-0.07)	-0.436 (-0.11)	-0.447 (-0.06)	0.245 (0.02)
D21, 22(S)	12.490 (0.14)	2.306 (0.12)	3.108 (0.09)	4.004 (0.07)
D21, 22(S^2)	-1.227 (-0.01)	-0.642 (-0.03)	-0.837 (-0.02)	1.590 (0.02)
D21, 22 (S^3)	-1.390 (-0.05)	-0.530 (-0.02)	-0.190 (-0.02)	-1.106 (-0.06)
D24, 26, 27	11.868 (0.20)	1.656 (0.13)	0.356 (0.02)	9.856 (0.25)
D24, 26, 27(S)	-43.935 (-0.22)	-4.518 (-0.10)	1.591 (0.02)	-41.008 (-0.30)
D24, 26, 27(S^2)	57.976 (0.34)	3.367 (0.09)	0.551 (0.01)	54.059 (0.46)
D24, 26, 27(S^3)	-14.090 (-0.37)	-0.695 (-0.08)	0.499 (0.03)	-13.894 (-0.54)

continued

TABLE 3.3 (continued)

Variable	Total R&D	Basic	Applied	Development
D28	−0.364	2.433	0.878	−3.536
	(−0.05)	(−1.52)	(0.31)	(−0.72)
D28(S)	37.881*	−6.488†	13.013†	30.608*
	(2.53)	(−1.94)	(2.22)	(3.01)
D28(S^2)	−9.906	4.636*	−2.646	−11.608*
	(−1.51)	(3.15)	(−1.03)	(−2.60)
D28(S^3)	1.218**	−0.512*	0.262	1.441*
	(1.69)	(−3.17)	(0.92)	(2.94)
D29	−2.120	1.043	−3.514	0.330
	(−0.09)	(0.22)	(−0.42)	(0.02)
D29(S)	1.129	−0.363	1.840	−0.065
	(0.16)	(−0.18)	(0.52)	(−0.01)
D29(S^2)	0.656	0.050	0.132	0.477
	(0.59)	(0.20)	(0.30)	(0.63)
D29(S^3)	−0.028	−0.001	−0.007	−0.019
	(−0.73)	(−0.13)	(−0.48)	(−0.75)
D30, 32	−10.322	−1.308	−2.353	−6.946
	(−0.66)	(−0.37)	(−0.38)	(−0.65)
D30, 32(S)	50.005	5.401	12.277	33.035
	(0.89)	(0.43)	(0.56)	(0.87)
D30, 32(S^2)	−8.695	−1.971	−2.466	−4.673
	(−0.23)	(−0.23)	(−0.16)	(−0.18)
D30, 32(S^3)	−0.513	0.190	0.004	−0.640
	(−0.07)	(0.12)	(0.01)	(−0.13)
D33, 34	0.091	1.195	0.223	0.147
	(0.01)	(0.45)	(0.05)	(0.02)
D33, 34(S)	11.560	7.477	2.468	7.415
	(0.47)	(1.32)	(0.25)	(0.43)

continued

TABLE 3.3 (continued)

Variable	Total R&D	Basic	Applied	Development
D33, 34 (S^2)	-0.667	-2.149	0.063	-1.623
	(-0.08)	(-1.22)	(0.02)	(-0.30)
D33, 34 (S^3)	-0.001)	0.159	-0.012	0.133
	(-0.01)	(1.17)	(-0.05)	(0.32)
D35	5.049	0.277	1.704	3.069
	(0.62)	(0.15)	(0.54)	(0.56)
D35(S)	-6.797	-1.096	-5.953	0.252
	(-0.29)	(-0.21)	(-0.65)	(0.02)
D35(S^2)	19.023	0.937	8.099	9.988
	(1.35)	(0.30)	(1.46)	(1.04)
D35(S^3)	-1.339	-0.084	-1.043	-0.212
	(-0.60)	(-0.17)	(-1.19)	(-0.14)
D36	-2.037	-0.341	-0.316	-1.380
	(-0.19)	(-0.14)	(-0.07)	(-0.19)
D36(S)	54.764	1.995	9.215	43.555**
	(1.54)	(0.25)	(0.66)	(1.81)
D36(S^2)	-23.415	-0.929	-4.125	-18.360
	(-1.23)	(-0.22)	(-0.55)	(-1.42)
D36(S^3)	1.376	0.053	0.257	1.066
	(1.28)	(0.22)	(0.61)	(1.45)
D37	-7.308	0.297	-2.859	-4.771
	(-0.86)	(0.16)	(-0.86)	(-0.83)
D37(S)	29.344*	-0.760	9.488	20.695*
	(3.03)	(-0.36)	(2.49)	(3.15)
D37(S^2)	-0.677	0.354	-0.534	-0.513
	(-0.33)	(0.76)	(-0.66)	(-0.36)

continued

TABLE 3.3 (continued)

Variable	Total R&D	Basic	Applied	Development
D37(S^3)	0.013	-0.007	0.008	0.012
	(0.33)	(-0.74)	(0.53)	(0.44)
D38	-5.656	-0.096	-1.472	-4.088
	(-0.18)	(-0.01)	(-0.12)	(-0.19)
D38(S)	58.548	1.451	12.864	44.233
	(0.31)	(0.03)	(0.17)	(0.34)
D38(S^2)	58.531	-1.909	10.436	50.004
	(0.30)	(-0.04)	(0.14)	(0.38)
D38(S^3)	-17.636	0.609	-3.070	-15.176
	(-0.44)	(0.07)	(-0.19)	(-0.55)
R^2	0.976	0.701	0.862	0.984
F-level	123.189	7.082	18.876	181.111

*Significant at .01 level.
†Significant at .05 level.
**Significant at .10 level.
Note: t-statistics are reported in parentheses below each estimated coefficient. D is a binary variable; for example, D20 = 1 for all firms in SIC 20 and D20 = 0 otherwise.

Union Carbide, and Procter and Gamble. This influence is most noticeable in the equations related to basic research and to development. Specifically, with respect to total R&D expenditures, intensity increases with firm size among the smallest firms (sales less than $1.0 billion); then it decreases slightly among medium-sized firms and, finally, increases among the largest firms. Regarding basic-research expenditures, intensity increases with firm size among the smallest and medium-sized firms (sales less than $3.0 billion), and decreases among the largest firms. The relationship between applied-research intensity and firm size is similar to that for total R&D expenditures and size. Finally, development intensity increases

FIGURE 3.5

Graphical Representation of Chemical-Industry Equations: Total R&D

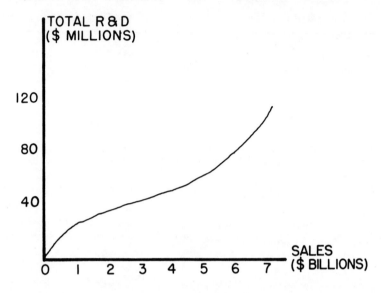

FIGURE 3.6

Graphical Representation of Chemical-Industry Equations: Basic

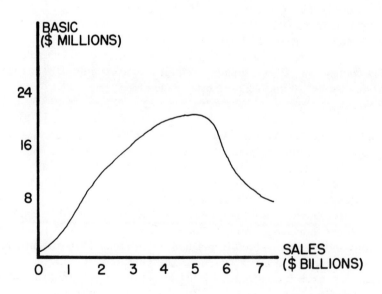

FIGURE 3.7

Graphical Representation of Chemical-Industry Equations: Applied

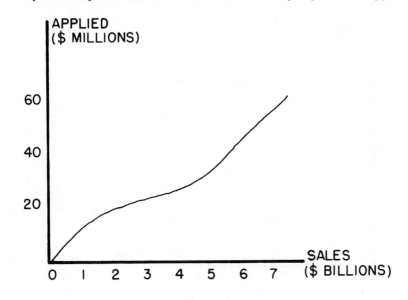

FIGURE 3.8

Graphical Representation of Chemical-Industry Equations: Development

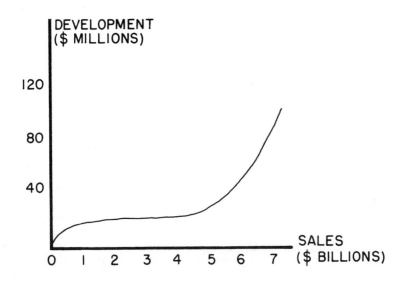

FIGURE 3.9

Graphical Representation of Machinery-Industry Equations: Total R&D

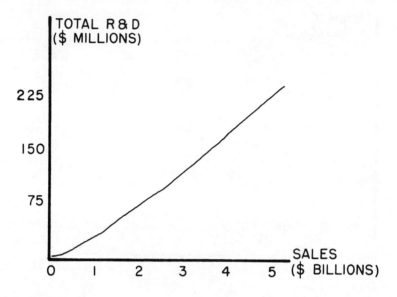

FIGURE 3.10

Graphical Representation of Machinery-Industry Equations: Basic

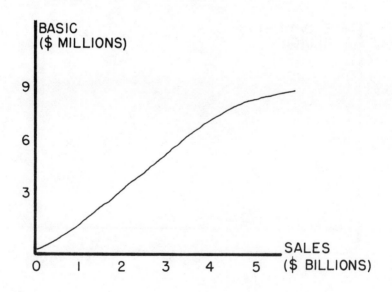

FIGURE 3.11

Graphical Representation of Machinery-Industry Equations: Applied

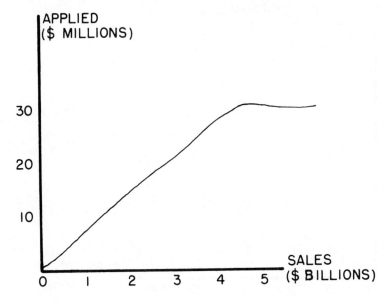

FIGURE 3.12

Graphical Representation of Machinery-Industry Equations: Development

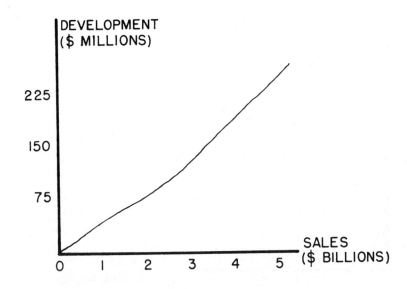

FIGURE 3.13

Graphical Representation of Transportation-Equipment-Industry Equations: Total R&D

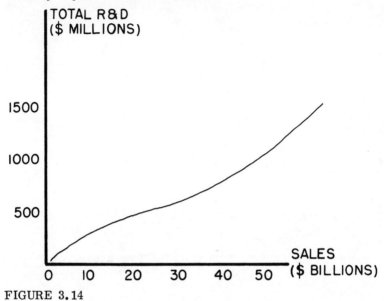

FIGURE 3.14

Graphical Representation of Transportation-Equipment-Industry Equations: Basic

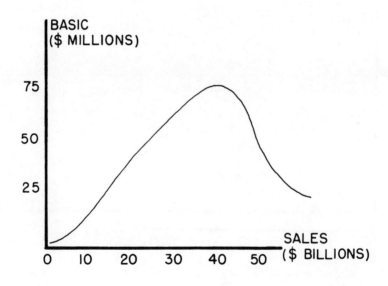

FIGURE 3.15

Graphical Representation of Transportation-Equipment-
Industry Equations: Applied

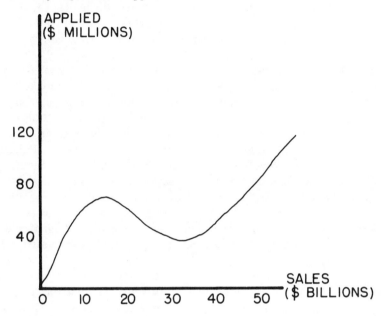

FIGURE 3.16

Graphical Representation of Transportation-Equipment-
Industry Equations: Development

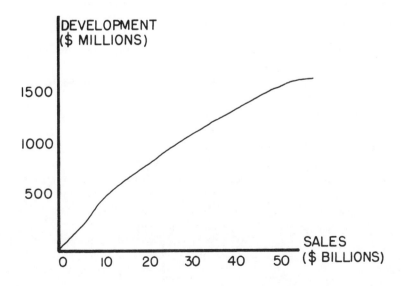

48 / FIRM SIZE AND INNOVATIVE ACTIVITY

among the very largest firms. Here, the importance of firm size over the widest range of firm sizes is especially evident with respect to basic research, and least evident with respect to development.

These empirical findings for the chemical industry contradict the findings of previous researchers who have concluded that R&D intensity increases, or at least remains constant, over all levels of firm size. One explanation is that the nature of R&D activity in the chemical industry has been changing over time. The earlier studies by Scherer (1965b) and by Mansfield (1968b) used R&D data pertaining to periods prior to 1960. There is empirical evidence that the costs and risks associated with R&D have increased greatly since the early 1960s. In particular, the 1962 amendments to the 1938 Federal Food, Drug, and Cosmetic Act have been shown to affect the nature of innovative behavior in the pharmaceutical industry (Peltzman 1973), whose firms comprise 16.0 percent of our sample of chemical firms. This event, and others, would suggest that only the largest firms would be able to provide the physical and financial economies of scale necessary for rigorous R&D activity. It may also suggest an emerging pattern of innovative leadership among chemical firms, under the assumption that input intensity is positively correlated with success in innovation.

The estimated regression results that correspond to the machinery industry (SIC 35) are graphically represented in Figures 3.9 through 3.12. In general, the graphs indicate that innovation intensity increases with firm size across firms of all sizes. There is some indication that applied-research intensity decreases in the very largest firms. Scherer (1965b), however, found that total R&D intensity increases with firm size only among the smallest and medium-sized firms. Our results are not strictly comparable with his since our subsample of machinery firms does not contain the largest firms in that industry: IBM, International Harvester, and Caterpillar Tractor, for example. In fact, IBM's sales are more than five times the sales of the largest firm in our subsample, Deere. Therefore, our results are biased and only represent the behavior of the smallest and medium-sized firms, and must be interpreted comparatively with that qualification.

The regression results corresponding to the transportation-equipment industry are shown in Figures 3.13 through 3.16. First, total R&D intensity increases with firm size only among the smallest firms (those with sales less than $5.0 billion), and then it remains fairly constant. Basic-research intensity increases with firm size among the smallest and medium-sized firms, (sales less than $35.0 billion), and then decreases only among the very largest firms. Similarly, applied-research intensity increases with size only among the smallest firms and among the very largest firms. The impact of General Motors dominates the observed behavior of the largest firms. Development intensity, like total R&D intensity, reaches a maximum at a size level of about $5.0 billion and then remains fairly constant over the remaining spectrum of firm sizes. Again, it is evident that firm size has the greatest impact over the widest range of firm sizes with regard to basic-research intensity, and the least impact with regard to development intensity.

CONCLUSIONS

As discussed above, the empirical results of studies like this may have definite policy implications for the structure of particular manufacturing industries and perhaps for the entire manufacturing sector of the economy. Our analysis of three leading R&D industries does not indicate that firm size per se is a prerequisite for innovation intensity. This conclusion is not unfamiliar; however, it could be extremely misleading. Our analysis indicates that firm size is relatively more important as a determinant of basic-research intensity than of either applied-research or development intensity. The fact that some 96 percent of all R&D is devoted to either applied or development activity overrides this basic-research-to-size relationship shown in most other aggregated empirical R&D studies.

We have casually contended that R&D activity is an input into the innovation process, and that innovation is a primary determinant of technological growth. Few economists would dispute this generalization. It may be argued, however, that basic research is the relatively more important input into the innovation process and into the advancement of technology. In fact, what is usually categorized as applied research or as development is activity generally devoted to product improvement or to product imitation. These latter two activities are not those responsible for shaping the structure of capitalism, and are obviously not those that Schumpeter envisioned as influencing the process of creative destruction. It is important, then, to realize that the effect of size is not constant across industries or across uses of R&D. Any policies attempting to argue that firm size is not a prerequisite for R&D spending must be specific as to the industry use of the R&D. A policy limiting firm growth or mergers may have adverse effects in stimulating the intensity of basic research, but may have no impact on the intensity of development activity. At least this is potentially the case in the chemical and transportation-equipment industries.

4 RATES OF RETURN TO R&D EXPENDITURES

Since the pioneering study by Solow (1957), where it was shown that 87 percent of the productivity increases in the U.S. economy between 1909 and 1949 were attributable to technological change, researchers have been interested in identifying the sources of technology that accounted for this growth, or that may account for future growth. It is well known that technology, in its broadest sense, results from the creation and distribution of technical knowledge. Technical knowledge, in turn, results from innovation, and one of the principal inputs into the innovation process is R&D activity. This is not to say that the results of R&D activities are the only input into innovation. The works of Jewkes, Sawers, and Stillerman (1969) and of Samuel Hollander (1965) clearly demonstrate that individual inventors have had a significant impact in formulating the direction of innovative activity. However, R&D activity is one specific input that can be easily identified and hence quantified.

Estimating the influence of R&D activity, R&D expenditures in particular, on technological growth was basically an academic question during the decade or so following Solow's findings. During that time, aggregate productivity was growing at a healthy pace, and policy makers had more pressing concerns than that of isolating target variables to be used for stimulating the economy. Today, however, this question is of practical importance. The rate of productivity growth—output per man-hour—has been declining throughout the economy, and no one seems to be able to target the influences that are responsible (Denison 1979). In the manufacturing sector, for example, the average annual rate of productivity growth between 1960 and 1965 was 3.4 percent; between 1965 and 1973, 2.8 percent; and between 1973 and 1977, 1.4 percent (Griliches 1979). Many suggest that this slowdown is related to the overall slowdown of technological growth in the economy, and hence they have been intent on examining the relationship between R&D activity and productivity growth, especially during the late 1970s. If in fact it can be shown that R&D-related innovations are a significant determinant of productivity growth, that may suggest additional policy avenues for stimulating the economy.

In this chapter, it is demonstrated empirically that there has been a collapse in the R&D-to-productivity-growth relationship, but only in the smaller manufacturing firms. The results of the analyses suggest that firm size is a prerequisite for successful R&D activity.

A REVIEW OF PREVIOUS STUDIES

The majority of researchers to date have concluded that productivity increases are positively related to investments in technology, i.e., R&D expenditures, and that the marginal return earned on these investments is very large. Some researchers have even concluded that these later findings suggest an underinvestment in R&D activity—it is interesting to note that A. C. Pigou (1932) reached a similar conclusion more than 50 years ago.

Before reviewing these studies, it may be useful to describe the model underlying the analyses. Usually, it is assumed that the firm is characterized by a three-factor production function (Y), written in terms of labor (L), physical capital (K), technical capital (T), and a disembodied technology coefficient (A):

$$Y = A\,F(L, K; T) \tag{4.1}$$

Productivity growth can mathematically be defined and measured as a Solow-type residual, $\dot{A}/A$:

$$\dot{A}/A = \dot{Y}/Y - \beta(\dot{L}/L) - (1 - \beta)(\dot{K}/K) = f(T), \tag{4.2}$$

where the dot notation refers to a time derivative, $\dot{Y} = dY/dt$, and where β is the output elasticity of L.

Jora Minasian (1962) was one of the first researchers to investigate the relationship between productivity and R&D expenditures. He examined residually measured productivity growth, from 1947 to 1957, in 18 firms from the chemical and allied-products industry and in five firms from the drug and pharmaceutical industry, and found there was a very high and statistically significant correlation between R&D expenditures and these residual measures. In addition, he showed that R&D expenditures were strongly related to both the level and the trend of profitability in firms, indicating "that there are gross returns to [R&D] expenditures" (Minasian, 1962, p. 141).

The first systematic attempt to estimate the marginal rate of return to R&D in the manufacturing sector was by Mansfield (1965). Using data taken from five chemical and five petroleum firms over the 1946-62 period, he concluded that R&D expenditures are significantly related to residually measured productivity, and that the marginal return on these investments is between 7 and 30 percent in the chemical firms, and between 40 and 60 percent in petroleum firms. Mansfield, however, did point out that the value of these returns is only suggestive since his underlying model was formulated on the basis of highly simplifying assumptions. Also, these rates do not reflect any interfirm or interindustry spillovers of R&D-related technical knowledge and thus are perhaps on the conservative side.

Minasian (1969) later found that the output elasticity of R&D expenditures was less than the output elasticity of physical capital; however, the gross private return on each of these investments was 54 percent and 9 percent, respectively. These conclusions came from estimating a log-linear version of a Cobb-Douglas production function similar to that described in equation 4.1, where technical capital (T) is approximated by the private R&D expenditures taken from each of 17 chemical firms. The sample period covered 1948 to 1957.

Nestor Terleckyj's (1974) analysis of the effects of R&D on the productivity growth (also defined as a Solow residual) of industry represents one of the most extensive investigations to date. His sample was limited to observations of two-digit and three-digit industries between 1948 and 1966. Of the 33 industries in the sample, 20 were classified as being in the manufacturing sector. The model studied was again derived from a production function similar to that described in equation 4.1. Terleckyj's final estimating equation related a measure of total factor productivity to alternative indexes of R&D intensity, ceteris paribus. The percentage of each industry's total sales to other industries, the percentage of unionization in each industry, and an index of the cyclical instability of each industry's output were the other variables held constant in his regressions. Terleckyj concluded that the marginal rate of return to self-financed R&D in manufacturing was about 30 percent, and that the rate of return to R&D investments embodied in inputs purchased was about 45 percent. The latter estimate is one of the first that explicitly attempts to quantify the interindustry spillover effects of R&D-related technology. Such a finding suggests that the social return associated with R&D is greater than the private return. Terleckyj viewed this finding as tentative support for justifying federally funded R&D in manufacturing—an issue to be discussed in more detail in Chapter 5.

Mansfield et al. (1977) conducted case studies of 17 selected innovations of manufacturing firms in an attempt to reinvestigate the returns to R&D investments at a microeconomic level. Generally, the innovations examined were first begun in the early or middle 1960s. Of these 17, 13 were product innovations and four were process innovations. Based on interviews and an analysis of restricted data, it was concluded that the median social return was about 56 percent and that the median private return was about 25 percent. These estimates compare favorably with those from Terleckyj's production-function study.

Zvi Griliches (1973, p. 80) avers that the estimates from these and other studies could be "too high perhaps by as much as 50 percent," not only because of the simplifying assumptions underlying the production-function models, but also because of the implicit assumption that the firms relevant stock of technical knowledge is proportional to the flow of R&D expenditures. As an attempt to avoid this problem, Albert Link (1978) formulated and estimated a model of induced technology. The model is based on the assumption that a firm spends money for R&D in order to finance technology that will lower its total costs of production. The empirical analysis, based on data from two-digit, three-digit, and four-digit manufacturing industries for 1958, revealed an average earned rate of return of about 19 percent—some 50 percent below the earlier estimates of 30 to 50 percent.

Link's estimate compares well with that of Griliches (1980), which was derived from a production-function model applied to a cross section of 883 large manufacturing firms (1,000-plus employees) on date from 1957 to 1965. He estimated that the overall return in manufacturing to company-financed R&D expenditures was also about 19 percent. There were, however, sizable interindustry differences: in the chemical and petroleum industries, the estimated marginal return was 103 percent, and it was only 3 percent in the electrical and electronic-equipment industries.

Three conclusions can be drawn from these studies: one, residually measured productivity growth is positively related to R&D expenditures both at the firm and at the industry levels; two, the private marginal rate of return to R&D expenditures is quite high, at least relative to the competitive return earned on physical capital; and, three, the social return from innovative R&D is greater than the private return. These conclusions must be tempered in that each has been generalized from data corresponding to periods from the late 1940s to the mid-1960s. It is unjustified—and recent empirical evidence supports this—to assume that the R&D-to-productivity relationship is as strong in the 1970s as it was in the late 1960s.

Griliches (1979) has recently asked whether the slowdown in productivity growth that began in the early 1970s could be explained by the slowdown in R&D expenditures (real and nominal) over that period. Although that conclusion is not probable, an interesting relationship is revealed by his analysis. He found, using productivity and R&D data from 39 industries at the three-digit level, that the return to productivity from R&D was zero over the period 1969 to 1977. A similar finding was also reported by Carson Agnew and Donald Wise (1978), who used a more aggregated sample of two- and three-digit industries. If these findings are interpreted literally, then, as Griliches (1979) suggests, the slowdown in productivity growth is partially a result of the collapse in the productivity of R&D. There are at least three reasons that partially explain this phenomenon: one, firms may be diverting R&D resources from their previous role as an input into innovation to a new role as an input into the less productive role of meeting regulatory constraints; two, the technical effectiveness of R&D, direct and indirect, is to some extent embodied in the newer vintages of physical capital, and the recent investment slowdown in manufacturing may simply be postponing the impact of those R&D expenditures on productivity growth; and, three, firms may be uncertain about future economic events and are therefore using R&D resources in a defensive manner, rather than in a more innovative fashion, that results in shifting their production frontiers.

In the analysis that follows, the returns to R&D are reestimated over the period 1971 to 1976, using firm data rather than industry aggregates. The purpose is to gain additional insights into the causes of the apparent collapse in R&D productivity.

THE ANALYTICAL FRAMEWORK

Here, the effect of R&D investments on productivity is examined. Residually measured productivity growth and R&D expenditures are estimated for each of the 174 firms in the sample-data set described in Chapter 2. The model is based on a Cobb-Douglas production function similar to that described by equation 4.1.

The Empirical Model

The model used here is formulated on the assumption that the firm operates according to a three-factor production function:

$$Y = A\ F(L, K; T), \tag{4.3}$$

where Y represents output, A is a neutral disembodied shift parameter, L and K are measures of the stock of labor and of physical capital, respectively, and T is the stock of technical capital or technical knowledge. T, in turn, is written as a function of specific technical capital (C), and of the other factors (O) that affect its production:

$$T = G(C, O), \qquad (4.4)$$

where research capital (C) is some weighted accumulation of previous R&D investments (R)

$$C = \Sigma a_i R_{t-i}. \qquad (4.5)$$

These accumulation weights (a_i) reflect, in theory, the influence of both a distributed lag effect of i periods on past R&D and the rate of obsolescence for research capital.

If the production function in equation 4.3 has the form of a Cobb-Douglas function, then the model becomes

$$Y = A_o e^{\lambda t} L^\beta K^{(1-\beta)} T^\alpha, \qquad (4.6)$$

where A_o is a constant, λ is a disembodied rate-of-growth parameter, and β and α are output elasticities. Constant returns to scale are assumed only with respect to labor (L) and capital (K).

Differentiating equation 4.6 with respect to time (t), it follows that

$$\dot{Y}/Y = \lambda + \beta(\dot{L}/L) + (1-\beta)(\dot{K}/K) + \alpha(\dot{T}/T), \qquad (4.7)$$

where dot notation refers to a time derivative, $\dot{Y} = dY/dt$. Residual productivity growth, as defined by equation 4.2 is

$$\dot{A}/A = \dot{Y}/Y - \beta(\dot{L}/L) - (1-\beta)(\dot{K}/K). \qquad (4.8)$$

Substituting $\dot{A}/A$ from equation 4.8 into a rearranged version of equation 4.7, equation 4.9 is obtained:

$$\dot{A}/A = \lambda + \alpha(\dot{T}/T) \qquad (4.9)$$

The parameter α is the output elasticity of technical capital:

$$\alpha = (\partial Y/\partial T)(T/Y). \qquad (4.10)$$

Substituting the right-hand side of equation 4.10 into equation 4.9, and rearranging, we have

$$\dot{A}/A = \lambda + \rho(I_T/Y), \qquad (4.11)$$

where $\rho = (\partial Y/\partial T)$ is interpreted as the marginal product of technical capital, and $I_T = \dot{T}$ is the net private investment of the firm in technical capital.

If equation 4.11 is stochastic, rather than deterministic, it can be written as

$$\dot{A}/A = \lambda + \rho(I_T/Y) + \epsilon, \qquad (4.12)$$

where ϵ is a random disturbance term. Versions of equation 4.12 have been used by other researchers, and one will be used in the following analyses. An empirical estimate of the slope coefficient ($\hat{\rho}$) will represent the marginal rate of return to R&D, assuming that per-period R&D expenditures represent the relevant net investments (I_T) in the firm's stock of technical capital.

Although this model is common among researchers in the area of productivity growth and R&D activity, several conceptual problems are inherent in its formulation. First, it is important to note that the model is not based on any assumed behaviors of the firm. Observed R&D expenditures are assumed to be the relevant (optimal) net investments in the stock of technical capital, and technical capital is postulated to augment productivity homogeneously across observations. In other words, R&D investments are implicitly a rational activity of the firm directed optimally toward technological growth. This may in fact be the case. At least, Schumpeter (1947) and Fritz Machlup (1962) promote the view that innovative activity is a rational process. There is an alternative view that should be acknowledged. Donald Schon (1967) and Bela Gold (1971) suggest that the process of innovation is sporadic and irregular, and consequently cannot be analyzed within the neoclassical confines of managerial choice.

An even more fundamental problem exists with respect to the accuracy with which $\dot{A}/A$ measures productivity growth. Whether output (Y) is measured as value added or as sales, it will not reflect time-related improvements in the quality of the good or service produced (Griliches 1973). To the extent that such quality improvements result from R&D investment, there will be a bias in the estimated value of $\dot{A}/A$ and hence in $\hat{\rho}$. Also, many firms sell their R&D output directly to the public sector—especially those involved in defense- or space-oriented R&D. As Griliches (1973) correctly notes, these products are measured in terms of their cost, and hence they definitionally do not show up in a residual calculation. Unfortunately, few solutions are available to resolve these problems. Perhaps the best alternative is to use the most disaggregated data available, and to investigate R&D not as a single homogeneous activity, but according to its character of use.

The Statistical Estimation of Rates of Return

Equation 4.12 was estimated using the entire sample of 174 manufacturing firms. $\dot{A}/A$ was computed for each firm as the average annual rate of productivity growth over the period 1971 to 1976:

$$\dot{A}/A = 1/5[(\ln Y_{76} - \ln Y_{71} - \ln PD)$$
$$- \beta (\ln L_{76} - \ln L_{71}) - (1 - \beta)(\ln K_{76} - \ln K_{71})]. \qquad (4.13)$$

The data for each of these firm calculations came primarily from the Compustat data file. An output measure comparable to value added was constructed from information reported on each firm's income statement. Specifically, output was calculated as net sales less the nonlabor costs of goods sold (such as taxes other than income taxes, maintenance and repairs, heat, light, power, transportation charges, insurance, licenses), and so on. The price deflator (PD) used was the wholesale price index (1971 = 100) that corresponded to the four-digit industry in which each firm was classified by Compustat. These indexes are published annually by the Bureau of Labor Statistics in <u>Wholesale Prices and Price Indices</u>.

The average share of labor in total output (β) over the period 1971 to 1976 was calculated as the average of each firm's total annual labor expenditures per unit of output. Since most firms do not report their total expenditures to Compustat, these data were approximated by the product of the average four-digit manufacturing-industry wage corresponding to each firm (as calculated from <u>The Census of Manufacturers</u>) and the total labor force employed in each firm. Labor-force estimates (L) are available from Compustat, and are defined as the number of employees reported to stockholders.

An estimate of each firm's physical capital stock (K) was obtained from Compustat, and represented the historic book value of net plant defined as gross plant—representing tangible fixed property, such as land, buildings, and equipment—less accumulated reserves for depreciation, depletion, and amortization. Capital's share is simply calculated as $(1 - \beta)$.

Theoretically, I_T represents the time derivative of the stock of technical capital (T); but since there is no estimate of either the lag between the firm's R&D expenditures and the resulting increment in T, or the rate of obsolescence for innovations and technical knowledge, any numerical estimate of I_T will be subject to error. Here, I_T is conventionally measured as the firm's total company-financed R&D expenditures in 1976, as reported by Compustat. Equation 4.12 was estimated using ordinary least-squares analysis. The sample consisted of all 174 firms. I_T was measured in four alternative ways:

TABLE 4.1

Estimated Rates of Return to R&D Expenditures by
Character of Use: Sample of 174 Firms

Variable	Total R&D	Basic	Applied	Development
Intercept	-0.151	-0.170*	-0.160	-0.139
	(-1.20)	(-1.88)	(-1.45)	(-1.13)
(I_T/Y)	-0.001	0.130	0.006	-0.007
	(-0.03)	(0.41)	(0.07)	(-0.17)
R^2	0.001	0.001	0.001	0.002
F-level	0.001	0.169	0.005	0.028

*Significant at the .10 level.
Note: t-statistics are reported in parentheses below each estimated coefficient.

by total company-financed R&D expenditures, company-financed basic-research expenditures, company-financed applied-research expenditures, and company-financed development expenditures. By definition, the sum of basic-research, applied-research, and development expenditures equals total R&D expenditures. The least-squares results are reported in Table 4.1. It is quite clear that the overall specifications of the model are poor, as evidenced by the F-statistics. As well, the estimated rate-of-return coefficients, for each of the four alternative measures of I_T, are not significantly different from zero. Given the previous interindustry research of Griliches (1980) and of Agnew and Wise (1979), over similar time periods, such results are not unexpected.

In an attempt to control for obvious interindustry differences in the character of R&D and its impact on productivity, the sample of 174 firms was again subdivided into the 11 industry groupings defined in Table 3.2.

Versions of equation 4.12 were estimated using ordinary least squares, allowing for each of 11 separate intercept terms and each of 11 separate slope terms (as defined by the categories in Table 3.2), by the use of binary and interaction-binary variables. As above, I_T

TABLE 4.2

Estimated Rates of Return to R&D Expenditures by Character of Use and by Industry Groupings: Sample of 174 Firms

Variable	Total R&D	Basic	Applied	Development
Intercept	-0.272	-0.412	-0.192	-0.187
	(-0.47)	(-1.30)	(-0.36)	(-0.36)
D21, 22	0.453	0.542	0.357	0.334
	(0.61)	(1.15)	(0.54)	(0.48)
D24, 26, 27	0.241	0.340	0.174	0.134
	(0.26)	(0.63)	(0.19)	(0.16)
D28	-0.321	0.150	-0.205	-0.243
	(-0.43)	(0.387)	(-0.34)	(-0.38)
D29	0.094	0.440	0.029	-0.029
	(0.10)	(0.84)	(0.03)	(-0.03)
D30, 32	0.019	0.476	0.057	0.188
	(0.02)	(1.03)	(0.08)	(0.26)
D33, 34	1.283	0.985	1.079	1.280
	(1.20)	(1.63)	(1.02)	(1.36)
D35	-0.229	0.021	-0.425	-0.183
	(-0.33)	(0.06)	(-0.70)	(-0.30)
D36	0.768	0.645	0.913	0.408
	(0.90)	(1.37)	(1.22)	(0.52)
D37	-0.453	-0.044	-0.238	-0.533
	(-0.63)	(-0.10)	(-0.37)	(-0.79)
D38	0.300	0.154	0.317	0.164
	(0.30)	(0.28)	(0.35)	(0.17)
(I_T/Y)	-0.049	0.574	-0.322	-0.197
	(-0.13)	(0.45)	(-0.33)	(-0.35)

continued

TABLE 4.2 (continued)

Variable	Total R&D	Basic	Applied	Development
D21, 22 · (I_T/Y)	−0.014 (−0.03)	−1.029 (−0.51)	−0.059 (−0.04)	0.141 (0.22)
D24, 26, 27 · (I_T/Y)	0.023 (0.05)	−0.639 (−0.34)	0.239 (0.18)	0.171 (0.23)
D28 · (I_T/Y)	0.115 (0.31)	−0.622 (−0.46)	0.383 (0.39)	0.259 (0.44)
D29 · (I_T/Y)	0.106 (0.16)	−3.168 (−0.59)	0.433 (0.30)	0.355 (0.33)
D30, 32 · (I_T/Y)	0.236 (0.58)	1.106 (0.59)	0.859 (0.77)	0.346 (0.57)
D33, 34 · (I_T/Y)	−0.209 (−0.39)	−0.982 (−0.49)	−0.358 (−0.21)	−0.327 (−0.40)
D35 · (I_T/Y)	0.101 (0.27)	0.960 (0.48)	0.810 (0.78)	0.215 (0.37)
D36 · (I_T/Y)	−0.097 (−0.26)	−0.973* (−2.45)	−0.740 (−0.67)	0.091 (0.16)
D37 · (I_T/Y)	0.201 (0.54)	2.914 (1.29)	0.601 (0.59)	0.406 (0.67)
D38 · (I_T/Y)	−0.013 (−0.03)	−1.355 (−0.39)	−0.098 (−0.09)	0.131 (0.22)
R^2	0.090	0.122	0.104	0.079
F-level	0.685	0.955	0.801	0.600

*Significant at the .05 level.

Note: t-statistics are reported in parentheses below each estimated coefficient. D is a binary variable; for example, D20 = 1 for all firms in SIC 20 and D20 = 0 otherwise.

was measured as total R&D expenditures and alternatively by each of the three categories of R&D use. The estimated results are shown in Table 4.2. Again, it is clear that none of the regression specifications is significant, and in no instance is the marginal rate-of-return coefficient positive and significantly different from zero.

Although it is unusual to present econometric results that are statistically insignificant as empirical support for a testable proposition, the regression results reported in Tables 4.1 and 4.2 substantiate the findings of others. In fact, these results may suggest that the collapse in the productivity of R&D is not specific to any one use of R&D expenditures, but is common to basic-research, applied-research and development expenditures.

Of the 11 industry groupings described in Table 3.2, three have a sufficient number of observations to allow for an individual econometric analysis: the chemical and allied-products industry (SIC 28), the machinery industry (SIC 35), and the transportation-equipment industry (SIC 37). These three industries are especially representative of industrial R&D. They represent three of the five leading R&D industries, and in 1976, their company-financed R&D amounted to over 55 percent of all company-financed R&D in the manufacturing sector. Versions of equation 4.12 were estimated using ordinary least squares. Again, four alternative measures of I_T were used in each of these industry models.

TABLE 4.3

Estimated Rates of Return to R&D Expenditures by Character of Use: Sample of 33 Chemical Firms

Variable	Total R&D	Basic	Applied	Development
Intercept	-0.593*	-0.263	-0.397	-0.430*
	(-3.68)	(-3.10)	(-3.56)	(-3.38)
(I_T/Y)	0.065†	-0.048	0.061	0.061
	(2.17)	(-0.30)	(1.42)	(1.46)
R^2	0.140	0.003	0.065	0.069
F-level	4.711	0.089	2.002	2.144

*Significant at the .01 level.
†Significant at the .05 level.
Note: t-statistics are reported in parentheses below each estimated coefficient.

TABLE 4.4

Estimated Rates of Return to R&D Expenditures by Character of Use: Sample of 34 Machinery Firms

Variable	Total R&D	Basic	Applied	Development
Intercept	-0.500 (-1.17)	-0.391† (-2.37)	-0.617* (-2.72)	-0.370 (-1.44)
(I_T/Y)	0.052 (0.73)	1.534 (1.23)	0.488** (1.75)	0.017 (0.24)
R^2	0.018	0.050	0.096	0.002
F-level	0.526	1.509	3.067	0.059

*Significant at the .01 level.
†Significant at the .05 level.
**Significant at the .10 level.
<u>Note</u>: t-statistics are reported in parentheses below each estimated coefficient.

TABLE 4.5

Estimated Rates of Return to R&D Expenditures by Character of Use: Sample of 19 Transportation-Equipment Firms

Variable	Total R&D	Basic	Applied	Development
Intercept	-0.725 (-0.76)	-0.457 (-0.68)	-0.430 (-0.51)	-0.721 (-0.76)
(I_T/Y)	0.152 (0.72)	3.488 (0.80)	0.279 (0.44)	0.209 (0.73)
R^2	0.30	0.037	0.011	0.031
F-level	0.525	0.648	0.192	0.537

The estimated results for the chemical industry are reported in Table 4.3. The estimated coefficient for total R&D expenditures is statistically significant at the .05 level, suggesting a rate of return of 6.5 percent. Examining the coefficients for basic-research, applied-research, and development expenditures, it may not be unreasonable to conclude that this estimated 6.5 percent return is reflecting the contribution of applied research and development relatively more than that of basic research, even though none of these individual coefficients is statistically significant.

The estimated results for the machinery industry are reported in Table 4.4. Only the coefficient for applied research is significant at the .10 level, suggesting a marginal rate of return of 48.8 percent. Still, these estimated results are tentative and at best suggestive.

The estimated results corresponding to the transportation-equipment industry are reported in Table 4.5. None of the specifications is significant, and virtually nothing can be inferred from the regression coefficients, other than that the estimated marginal returns to all forms of R&D are zero.

In summary, the results of these tests are similar to the findings of previous researchers.

A REFORMULATION OF THE SCHUMPETER HYPOTHESIS

In Chapter 3, Schumpeter's hypothesis that innovative activity is an increasing function of firm size was tested empirically under the conventional assumption that R&D expenditures are proportional to successful innovative activity.

Here, that proportionality assumption is relaxed, and innovative activity is viewed directly as an aspect of entrepreneurship. Modern studies of economics stress the omission of the entrepreneur in the standard neoclassical equilibrium analysis, and consequently have fostered new approaches toward understanding the concept and role of entrepreneurship. For example, Israel Kirzner (1973, p. 35) views entrepreneurship as a firm characteristic depictable as the managerial ability to perceive and adjust to a disequilibrium situation:

> Now I choose to label that element of alertness to possibly new worthwhile goals and to possibly newly available resources—which [is] absent from the notion of economizing but very much present in that of human action—the <u>entrepreneurial</u> element in human decision making.

Theodore Schultz (1975, p. 832) has even broadened the definition of entrepreneur to include all behavior-optimizing individuals:

Whether or not economic growth is deemed to be "progress," it is a process beset with various classes of disequilibria. In response, individuals in many different walks of life engage in optimizing behavior, which entails reallocating resources to regain equilibrium. All of them are in this respect entrepreneurs.

These modern views of the entrepreneur as an individual perceiving disequilibria, and then adjusting to a new equilibrium, are in contrast to the Schumpeterian notion that the entrepreneur is an individual who creates a disequilibrium situation from an equilibrium state within the overall circular flow process. Although these differences are quite distinct, and charge the entrepreneur with different responsibilities, the concept that entrepreneurial innovation is a process is fundamental. Whenever a process notion is considered, a question arises as to what is the best criterion for evaluating its success. It is suggested here that the rate of return earned on R&D expenditures is conceptually more appropriate as a measure of successful innovative activity than is some absolute index like R&D expenditures. Thus, the following section formulates an empirical test of Schumpeter by examining whether the rate of return to R&D is a function of firm size.

Structural Change in the Rate-of-Return Equation

Recall equation 4.12, which was the basis for the above estimation of the rate of return to R&D expenditures:

$$\dot{A}/A = \lambda + \rho(I_T/Y) + \epsilon. \tag{4.14}$$

Given a cross-sectional sample of firms, $\hat{\rho}$ estimates the sample's average rate of return. Presumably, if there were an a priori definition of which firms in the sample were large and which were small (in the Schumpeterian sense), then the sample could be subdivided into these two regimes, using a binary or interaction-binary variable, and ρ could be estimated for each regime. The null hypothesis would be that the rate of return in large firms would be greater than that in small firms. Unfortunately, no such workable definition of alternative size regimes is known.

Alternatively, the statistical test for structural stability developed by R. L. Brown, J. Durbin, and J. M. Evans (1975) is employed to determine the dividing point for these regimes. If the structure of equation 4.14 changes over the ranks of the regressors by an index of firm size, this will result in a shift of the residuals when compared to a model assuming constant coefficients. The test statistic (S_r) is based on the normalized cumulative sum of squared residuals from

FIGURE 4.1

Forward and Backward Plots of Recursive Residuals:
Chemical, Machinery, Transportation Equipment Industries

Forward Plots

Chemical:

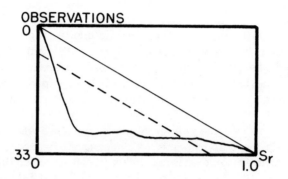

Machinery:

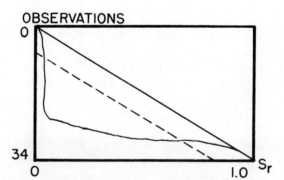

Transportation Equipment:

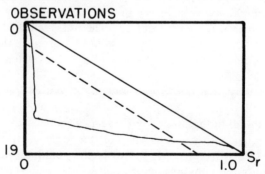

FIGURE 4.1 (continued)

Backward Plots

Chemical:

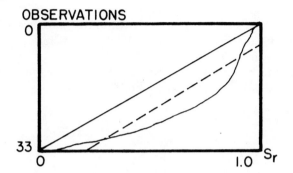

Machinery:

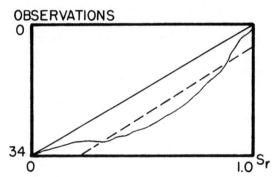

Transportation Equipment:

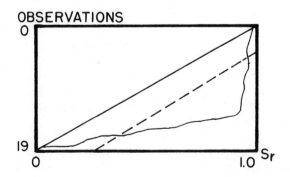

a recursive estimation model:

$$S_r = \left[\sum_{k+1}^{r} w_i^2 / \sum_{k+1}^{N} w_i^2 \right]; \quad r = k+1 \ldots N, \quad (4.16)$$

where w_i are the orthogonalized recursive residuals, k is the number of independent variables in the equation, and N is the number of behavior observations. S_r follows a beta distribution with a mean value $(r - k)/(N - k)$. If the regression coefficients are constant, a plot of S_r will lie along its mean-value line within the confidence limits $\{\pm C_o + [(r - k)/(N - k)]\}$ defined by Pyke's modified Kolmogorov-Smirnov statistic (C_o).

Regarding equation 4.14, the null hypothesis is that its structure is constant over all levels (1 ... N) of firm size:

$$H_o: \quad \begin{aligned} \lambda_1 &= \lambda_2 = \ldots = \lambda_N = \lambda \\ \rho_1 &= \rho_2 = \ldots = \rho_N = \rho \\ \sigma_1^2 &= \sigma_2^2 = \ldots = \sigma_N^2 = \sigma^2, \end{aligned} \quad (4.16)$$

with σ_i^2 (i = 1 ... N) being the variance of the error term (ϵ), in equation 4.14. The alternative hypothesis is that the estimated slope coefficient increases over alternative size regimes.

The Brown-Durbin-Evans test was applied individually to the sample of chemical, machinery, and transportation-equipment firms. For example, the forward plots of the S_r's against the observation number of (I_T/Y), where I_T is measured as total R&D expenditures, ranked in ascending order of firm size (measured by the firm's net sales), are shown in Figure 4.1. The mean-value line of S_r is shown as a solid diagonal, and the 99 percent confidence limit is shown as dotted lines. The departure of S_r from its mean denotes the observation where the structural shift becomes significant. It does not necessarily denote the observation where the shift initially began, or that only one shift occurred. Clearly, the null hypothesis of structural stability can be rejected at the 99 percent level of confidence or better. Examination of the backward plots shown in Figure 4.1 leads to the same conclusion. The Brown-Durbin-Evans test was also applied to variations of equation 4.14 for each alternative measure of I_T for the three industry samples. In every case, the null hypothesis of structural stability was rejected at the 99 percent level of confidence or better. The forward and backward plots are similar to those in Figure 4.1.

The Brown-Durbin-Evans test for structural stability has been applied in several other studies (Khan [1974]; Hodgson and Holmes [1977]; Heller and Khan [1979]; Stern, Baum, and Greene [1979]; and Link [1980]). Whenever the structure examined is determined to be unstable, there remains a methodological problem of deciding the

TABLE 4.6

Size Levels for Dividing Each Sample into Behavorial Regimes

Size Range	Threshold Size
Chemical Industry	
95.26-6,512.73	1,661.51
Machinery Industry	
95.01-4,403.89	1,261.00
Transportation-Equipment Industry	
201.49-47,181.00	3,918.53

Note: Size is measured as 1976 net sales in millions of dollars.

observation for dividing the data. There is precedent, once the assumption of stability has been rejected, for using Richard Quandt's (1960) log-likelihood-ratio test to pinpoint the observation to use in subdividing the data. Implicit in this methodology is the assumption of two behavioral regimes, although Link (1980) has generalized and applied Quandt's test to three regimes.

Quandt's test is used here to divide each sample of data into two alternative regimes. The size levels corresponding to these divisions are shown in Table 4.6, along with the range of firm sizes. In each industry, the threshold size for dividing the data was the same regardless of the way I_T was measured in equation 4.14. It should also be noted that, in the chemical and machinery industries, the threshold-size level corresponds to an intermediate-sized firm; in the transportation-equipment industry, the threshold-size level is relatively large.

Estimation of Rates of Return in Alternative-Sized Firms

A variation of equation 4.14 was used with these points of division to estimate the rate of return to R&D, by its character of use, in variously sized firms. Specifically, the following equation was estimated using ordinary least-squares analysis:

$$\dot{A}/A = \lambda_1 + \lambda_2 D + \rho_1 (I_T/Y) + \rho_2 D \cdot (I_T/Y) + \epsilon, \qquad (4.17)$$

where D is a binary variable equaling zero for firms with net sales less than or equal to the threshold-size level defined in Table 4.6, and equaling one for those with net sales greater than the threshold level. For the relatively smaller firms (D = 0), the estimated rate of return will be equal to $\hat{\rho}_1$; for the relatively larger firms, it will be $(\hat{\rho}_1 + \hat{\rho}_2)$.

The estimated regression results from equation 4.17 using the sample of chemical firms, are shown in Table 4.7. Each of the four specifications examined is highly significant, and the reported R^2 values are high by conventional standards. In the smaller firms, the estimated rate of return to total R&D expenditures is 27.7 percent; to basic research, 78.3 percent; to applied research, 54.7 percent; and to development, 19.2 percent. Each of these latter estimates is significantly different from zero at the .05 level or better.

The estimated results for the machinery industry are presented in Table 4.8. Each of the specifications is significant, as was the

TABLE 4.7

Estimated Rates of Return to R&D Expenditures by Character of Use and by Size of Firm: Sample of 33 Chemical Firms

Variable	Total R&D	Basic	Applied	Development
Intercept	-0.311	-0.148*	-0.221*	-0.234*
	(-3.57)	(-2.48)	(-2.77)	(-3.63)
D·intercept	-1.889*	-1.114*	-1.525*	-1.956*
	(-7.77)	(-6.50)	(-4.32)	(-8.54)
(I_T/Y)	0.030**	-0.078	0.027	0.027
	(1.85)	(-0.51)	(0.92)	(1.30)
$D·(I_T/Y)$	0.247*	0.861†	0.520†	0.175†
	(4.82)	(2.14)	(2.14)	(5.40)
R^2	0.798	0.623	0.615	0.799
F-level	35.617	14.841	14.350	35.818

*Significant at the .01 level.
†Significant at the .05 level.
**Significant at the .10 level.

Note: t-statistics are reported in parentheses below each estimated coefficient.

TABLE 4.8

Estimated Rates of Return to R&D Expenditures by Character of Use and by Size of Firm: Sample of 34 Machinery Firms

Variables	Total R&D	Basic	Applied	Development
Intercept	-0.108	-0.121	-0.037	-0.111
	(-0.46)	(-0.68)	(-0.22)	(-0.44)
D·intercept	-3.243*	-0.984*	-2.289*	-1.743†
	(-4.93)	(-3.01)	(-6.84)	(-2.47)
(I_T/Y)	0.003	0.561	-0.141	0.005
	(0.05)	(0.45)	(-0.50)	(0.68)
$D \cdot (I_T/Y)$	0.580*	5.857†	1.895*	0.315**
	(4.03)	(2.11)	(4.91)	(1.69)
R^2	0.499	0.300	0.670	0.221
F-level	8.958	3.865	18.284	2.553

*Significant at the .01 level.
†Significant at the .05 level.
**Significant at the .10 level.

case above, and the estimated rate of return to each form of R&D is not significantly different from zero. In the larger firms, the estimated return to total R&D spending is 58.3 percent (significant at the .01 level), while the return to basic research is 641.8 percent (significant at the .05 level); for applied research it is 175.4 percent (significant at the .01 level); and for development, it is about 32.0 percent (significant at the .10 level).

The estimated results for the transportation-equipment industry are shown in Table 4.9. The significance of the specifications is weaker than for the previous two cases, due to the decreased degrees of freedom in the estimation. As was the case for the machinery industry, the estimated rate of return in the smaller firms is not significantly different from zero. In the larger firms, the estimated rate of return to total R&D expenditures is 151.1 percent (significant at the .01 level); it is 692.7 percent for basic research (significant at the .10 level), 359.6 percent for applied research (significant at the .10 level), and 80.0 percent for development (significant at the .01 level).

TABLE 4.9

Estimated Rates of Return to R&D Expenditures by Character of Use and by Size of Firm: Sample of 19 Transportation-Equipment Firms

Variables	Total R&D	Basic	Applied	Development
Intercept	0.18	0.013	0.015	0.201
	(0.14)	(0.019)	(0.02)	(0.24)
D·intercept	-7.815*	-3.360**	-3.643	-7.40*
	(-3.12)	(-1.76)	(-1.66)	(-3.05)
(I_T/Y)	-0.066	0.561	-0.109	-0.137
	(-0.34)	(0.95)	(-0.17)	(-0.47)
$D \cdot (I_T/Y)$	1.577*	6.266**	3.705**	0.937*
	(3.19)	(1.73)	(1.85)	(3.09)
R^2	0.429	0.235	0.197	0.418
F-level	3.749	1.538	1.228	3.584

*Significant at the .01 level.
†Significant at the .05 level.
**Significant at the .10 level.

CONCLUSIONS

As the results presented above indicate, there is evidence that there has been a collapse in the productivity effects of R&D, totally and by character of use, in the manufacturing sector as a whole. Even when the chemical, machinery, and transportation-equipment industries were examined individually, this conclusion remained. However, with respect to these three industries (and perhaps for all of manufacturing), this collapse is specific to the smaller firms. In what has been defined as the smaller firms, the estimated rate-of-return coefficient is not statistically different from zero; this was not the case, however, in the larger firms. As discussed in Chapter 3, economies of scale (physical and financial) are an important determinant of R&D activity. It seems that, during years of productivity slowdown, only the larger firms had the capabilities to continue to perform productive R&D vis-à-vis defensive R&D, usually of a product-imitating nature.

From a policy point of view, R&D can continue to be viewed as a viable input into the process of productivity growth. In doing so, it must be noted that the impact varies across industries, and that different uses of R&D have differential impacts on productivity. In each of the three industries studied here individually, the magnitude of this relationship was greatest for basic research and least for development. This is to be expected—the risk and uncertainty associated with each activity is different. Quite logically, basic research is of a more long-term nature and, by definition, an exploration into the unknown. Accordingly, its marginal return is apt to be the greatest.

5 FEDERAL SUPPORT OF INDUSTRIAL R&D

The federal government is actively involved in supporting R&D activity throughout the economy. The U.S. Navy's sponsored research programs can be traced as far back as 1789; and the Department of Agriculture's involvement in the land-grant-college system dates from the mid-1800s (Markham 1962). Since World War II, federal support of R&D activity has increased dramatically, primarily in response to the nation's awareness of the need for military preparedness. This growth was aided by the establishment of a formal government contracting policy, and by the National Science Foundation Act of 1947 (Danhoff 1968). In Table 5.1, the percentage of the federal budget allocated to total R&D activities is shown for fiscal years 1940/79. The post-World War II growth is evident; the percentage allocated to R&D reached a high of 12.6 percent in 1965. The subsequent decline in that percentage is consistent with the decline in R&D spending discussed in Chapter 2.

In 1979, most of the federal R&D budget was allocated to industrial firms (52 percent), with the remainder divided between federal intramural programs (25 percent), universities and colleges (13 percent) and their federally funded research and development centers (5 percent), and other nonprofit organizations (5 percent). For the entire economy, 63 percent of total federal funds was directed to development, 24 percent to applied research, and 13 percent to basic research. As shown in Table 5.2, these distributions change across the three groups of performers. While industrial firms are relatively more development intensive, universities and colleges are more basic-research intensive.

Although Congress and the Executive Branch of the government ultimately determine the size of the government's total R&D obligations, the allocation of these funds is the responsibility of four

TABLE 5.1

Percent of Total Federal Budget Allocated to R&D Activities

Fiscal Year	Percent
1940	0.8
1941	1.4
1942	0.8
1943	0.8
1944	1.5
1945	1.7
1946	1.5
1947	2.4
1948	2.3
1949	2.7
1950	2.5
1951	2.8
1952	2.7
1953	4.0
1954	4.4
1955	4.8
1956	4.9
1957	5.8
1958	6.0
1959	6.3
1960	8.4
1961	9.5
1962	9.7
1963	10.8
1964	12.4
1965	12.6
1966	11.9
1967	10.7
1968	9.5
1969	8.9
1970	8.0
1971	7.6
1972	7.2
1973	7.1
1974	6.8
1975	6.0
1976	5.8
1977	5.8
1978 (est.)	5.7
1979 (est.)	5.6

Source: National Science Foundation 1979a.

TABLE 5.2

Federal R&D Budget by Performer
and by Character of Use

Performer and Character of Use	Percent of R&D Budget
Industrial firms	
Development	84
Applied research	14
Basic research	2
Federal intramural programs	
Development	51
Applied research	34
Basic research	15
Universities and colleges	
Development	12
Applied research	42
Basic research	46

Source: National Science Foundation 1979a.

principal agencies: the Department of Defense, the Department of Energy, the National Aeronautics and Space Administration (NASA), and the Department of Education. In 1979, the Department of Defense was responsible for the distribution of 46 percent of total federal R&D expenditures—more than three times that of any other agency (National Science Foundation 1979a).

It is obvious that the federal government's role in sponsoring R&D is well established, perhaps even a fact of life. Still, if for no other reason than academic purity, two important questions should be raised: why does the federal government support R&D (industrial, in particular), and, if support is justifiable (from an economic point of view), how should it be given?

A RATIONALE FOR FEDERALLY SUPPORTED R&D

The focus of this section is specifically limited to industrial R&D, although an evaluation of funds allocated to other performers is equally important, especially of funds to university R&D, which presumably results in pure knowledge—a rather intangible commodity. Nominally, the federal government justifies expenditures for industrial R&D on three grounds. First, the government is the sole producer of many goods and services with public-good characteristics; therefore, the government should maintain responsibility for providing financial support for technological growth among producers of these goods and services. The most noteworthy of these responsibilities are in national defense and space exploration. Other activities include air-traffic control, regulatory standards, and sewage disposal. Since the government is the direct purchaser of goods and services in these areas, it is argued that it can most effectively evaluate the relevant R&D needs (Committee for Economic Development 1980). Second, federal investments in R&D presumably correct for market failures in the production of products (such as energy) where the social returns from technological advancement are perceived to be greater than the private returns. Finally, the federal government is committed to the general advancement of technical knowledge and thus supports industrial R&D when the private level is less than socially optimal—that is, where there are positive externalities associated with the creation and distribution of technical knowledge (Mansfield 1976).

These justifications, although somewhat politically motivated, are not without economic foundation. The orthodox view in the theory of public finance dictates that federal government intervention is justifiable whenever the private market fails to allocate resources in a Pareto-efficient manner (Brumm and Hemphill 1976). Arrow (1962) has made what is perhaps the strongest case for the federal government's participation in the industrial-innovation process. He views innovation as a process of creating information, although information of a particular type and with a particular use, and he claims that information is a public good; that is, information possesses the characteristics of indivisibility and hence inappropriability. As well, the external benefits from information, presumably from its use, are positive. According to Arrow (1962, pp. 616-17), the social implications are obvious:

> Information is a commodity with peculiar attributes, particularly embarrassing for the achievement of optimal allocation. In the first place, any information obtained, say a new method of production, should, from the welfare point of view, be available free of

charge (apart from the cost of transmitting information). This insures optimal utilization of the information but of course provides no incentive for investment in research. In an ideal socialist economy, the reward for invention would be completely separated from any charge to the users of the information. In a free enterprise economy, inventive activity is supported by using the invention to create property rights; precisely to the extent that it is successful, there is an underutilization of the information.

Arrow's point is clear. And since private firms cannot fully internalize all the benefits from technical knowledge, or information, then it follows that they cannot fully appropriate the property rights; an underinvestment, at least from a social point of view, may result.

An underinvestment in R&D may also be predicted on the basis of uncertainty, and the inability of firms to adequately ensure against the failure of R&D projects. As discussed in Chapter 1, innovative (R&D-related) activity is characterized by both risk and uncertainty. An environment is created in which entrepreneurs must make decisions about the allocation of investment dollars to R&D and to other forms of investments, with imperfect information. Usually, the information required for such an allocative decision is relatively more imperfect with respect to the discounted rewards from R&D, since many of the parameters associated with success are simply of a random nature. Consequently, an individual entrepreneur will rationally avoid making an excessive commitment of resources to R&D. However, what he perceives as excessive, society may view as insufficient, especially for basic-research dollars.

Empirical researchers have been cautious in asserting that there exists an underinvestment in R&D in the private sector. There is substantial empirical evidence that the private marginal rate of return to R&D is relatively high; however, these estimates are not without error. Nevertheless, as Mansfield (1972) suggests, many empirical researchers conclude (with varying degrees of confidence) than an underinvestment may exist.

If in fact the private level of investment in R&D is less than the desired social level, it should be asked how, and to what extent, government participation should occur. The theory of welfare economics contends that it is efficient for the federal government to invest in industrial R&D to the point where the marginal social benefits derived from such investments equal the marginal social costs incurred. It goes without saying that the practicality of such a decision rule for use by public planners is wanting. Harold Brumm and John Hemphill (1976) correctly note that the applicability problems associated with the Pigouvian theorem of allocation do not necessarily imply that it is of no use for policy making. Rather, as they see it (1976, p. 37), "[in] lieu of hard empirical data on such [social] costs and benefits, the policy-maker can at least evaluate them qualitatively."

Some empirical evidence suggests that if policy makers are in fact using this Pigouvian allocation rule, at least qualitatively, they are not doing a very good job of it. It has been argued by researchers that there is a second determinant of the degree of federal R&D support for industry, which competes with Pigouvian efficiency; namely, if bureaucrats seek to maximize their own utility, the allocation of federal R&D dollars to industry may simply be a self-serving mechanism. Link (1977) has shown that the observed distribution of federal R&D dollars across two-digit industries is not inconsistent with such a self-interest hypothesis. Also, Link and Morrell (1980) have given empirical evidence that the interstate distribution of federal R&D obligations is, in part, politically motivated. Namely, it reflects the majority party's evaluation of states where the probability of obtaining additional legislative control is greatest.

THE SOCIAL GAINS FROM FEDERALLY SPONSORED R&D

Desirably, there should be direct as well as indirect positive social benefits from federally sponsored R&D projects. The direct benefits accrue to the firm in the form of increased productivity. Also, there will be the benefits of an increase in the scale of the firm's overall R&D program and in the human capital of its R&D scientists. For example, if a firm contracts to produce a given item of military hardware, the overall size of the firm's R&D establishment will probably increase. Morton Kamien and Nancy Schwartz (1975) suggest that there are economies of scale in R&D to be captured by the firm that could lower its overall average R&D costs and increase its average efficiency per R&D input. As well, technical knowledge will be gained by R&D scientists working on the contract project. Such additional knowledge will, to some extent, increase their share of human capital and hence their productivity in future company-related R&D projects.

The indirect social benefits associated with federally sponsored R&D relate to the extent to which the technology produced from it can be transferred to other firms in an industry or to other industries in general. This phenomenon is often referred to as a spillover effect. Unfortunately, there is strong opinion that this process of technological transfer is not working.

William Quesenberry (1979) points out that Americans are presently being taxed about $24 billion to support government sponsored

R&D, most of which is oriented toward the defense and space programs. But the technology generated is not being transferred toward commercial applications and hence is being underutilized. He lists at least three reasons for this. First, many entrepreneurs are unaware of the fact that military-sponsored technology is available for nongovernment use. There are certain available channels, but they appear rather inefficient. Although Congress is strongly supportive of technology transfer, Stephen Merrill (1979) suggests that legislative judgments as to how this should be accomplished are unfortunately, but perhaps, inevitably, nothing more than guesswork. According to Leonard Ault (1979), NASA's Technology Utilization Program (begun in 1962) may be the most extensive technology-transfer program in operation. Its purpose is to stimulate the commercialization of aerospace-development technology, which is encouraged primarily through publications, such as the <u>Patent Abstract Bulletin.</u> The program has been effective, at least in the engineering sciences. A second reason for the underutilization of existing technical knowledge is that successful technology transfer depends heavily on the availability of expert technical assistance. In most instances, however, little assistance is freely available. A third reason is that risk capital is not readily available to entrepreneurs who are willing to attempt utilization of this technology. Since there are sizable costs involved in bringing technology into the marketplace, even after it has been produced, the incentive placed before the entrepreneur for using this government-produced knowledge is small and perhaps even negative. Venture capital is not available, nor is this government knowledge available on an exclusive basis. Just as private firms have an incentive to underinvest in R&D since they cannot fully apportion the property rights to technical knowledge, so too they have an incentive to underinvest in transfers of technical knowledge.

Steps are being taken to stimulate economic productivity through federal R&D programs. In May 1978, President Carter created the President's Domestic Policy Review on Industrial Innovation, primarily in response to the decline, or the decay, according to some (Sheils et al. 1979), in domestic innovative activity over the past several years. The final report of this commission on the state of domestic innovative activity has led the President to induce stimulating measures. In his October 31, 1979, message to Congress, President Carter outlined nine initiatives for fiscal year 1981, to accomplish this goal. His first initiative was related to the transfer of technical information, as noted by Schlie (1979):

> Often, the information that underlies a technological advance is not known to companies capable of commercially developing that advance. I [President

Carter] am therefore taking several actions to ease and encourage the flow of technical knowledge and information. These actions include establishing the Center for the Utilization of Federal Technology at the National Technical Information Service, to improve the transfer of knowledge from Federal laboratories; and, through the State and Commerce Departments, increasing the availability of technical information developed in foreign countries.

There is very little empirical evidence on which to evaluate the direct social benefits of federal R&D, relative to increasing economic-productivity growth, and available evidence is mixed. The earliest studies evaluated the effectiveness of federally sponsored agriculture research. Griliches (1958), for example, estimated that as of 1955, the return to each dollar invested in hybrid corn was over 700 percent. For agriculture research in general, the return was more than 300 percent (Griliches 1964).

Terleckyj developed a model similar to that developed in equation 4.12; however, he included inter alia an independent variable measuring the intensity of government-sponsored R&D investments in technical capital:

$$\dot{A}/A = \lambda + \rho(I_T^C/Y) + \gamma(I_T^G/Y) + \Sigma_i \beta_i X_i + \epsilon. \qquad (5.1)$$

Recall that $\dot{A}/A$ is an index of residually measured productivity growth, where superscripts C and G refer to either company or government sources of funding technical-capital investments (I_T), and X represents other variables held constant. His data represented observations of 20 manufacturing industries. Terleckyj attempted to measure I_T^G in two alternative ways in order to test for either a direct or an embodied effect of federal intervention. First, I_T^G was measured as the average dollar value of government R&D funds allocated directly to each industry; and, second, it was measured as government-financed R&D attributed to each industry through the purchases of inputs from R&D-intensive industries. His empirical estimate of the return to either form of government-sponsored R&D ($\hat{\gamma}$) was zero. Agnew and Wise (1978) reached a similar conclusion by analyzing comparable manufacturing data over the period 1957 to 1975. Lawrence Goldberg (1978), using an entirely different model specification, corroborated the conclusion that there has been a minimal contribution to productivity from federal investments in R&D.

ALTERNATIVE ESTIMATES OF THE RETURN FROM FEDERAL R&D PROJECTS

It is not surprising that Terleckyj, Agnew and Wise, and Goldberg, among others, found a zero spillover effect associated with government R&D projects, given the paucity of available channels for technology transfer. But the finding that the direct return to the firm was also zero is somewhat puzzling. Paul Kochanowski and Henry Hertzfeld (1980) offer one explanation. The production-function framework common to Terleckyj's study, for example, assumes that all R&D is directed to process augmentation. Many of the government's efforts have been aimed at new products and hence will not show up empirically.

It is not clear whether innovation resulting from federal R&D contracts will directly affect the firm's residual productivity, especially when accepted accounting procedures require that government-contracted output be imputed to the firm at a value equal to its cost (Griliches 1973). Instead, if firms conducting federally sponsored research are able to internalize the benefits resulting from an increase in the scale of R&D activity and are able to extract from labor learned technical skills, then the effect of federal R&D would be to increase the efficiency of the firm's own R&D.

If the firm's total R&D budget is viewed as being determined by equating the expected marginal returns from R&D with the marginal costs of financing R&D, then increased R&D efficiency would reduce the effective cost of financing, and consequently expand the firm's R&D program downward along its marginal rate-of-return schedule. Consider the firm illustrated in Figure 5.1. Originally the firm is spending RD* dollars and earning, at the margin, a return equal to ρ (as estimated in Chapter 4). If participation in federal R&D projects lowers the effective marginal cost of conducting R&D, as hypothesized here, the cost schedule will shift rightward, reflecting both an increase in the firm's investments in R&D and a decrease in the marginal return, ceteris paribus. The magnitude of these changes are, of course, functions of the size and type of the government contract, and of the slope of the curves shown in Figure 5.1.

This effect can be estimated empirically by using the model developed in Chapter 4 in equations 4.3 through 4.11. Equation 4.12 is modified to allow the rate of return earned in privately financed R&D to be a function of the presence of federal R&D financial support as:

$$\dot{A}/A = \lambda + \rho_1 (I_T/Y) + \rho_2 \text{FRD} \cdot (I_T/Y) + \epsilon, \tag{5.2}$$

FIGURE 5.1

Optimal Level of Company-Financed R&D

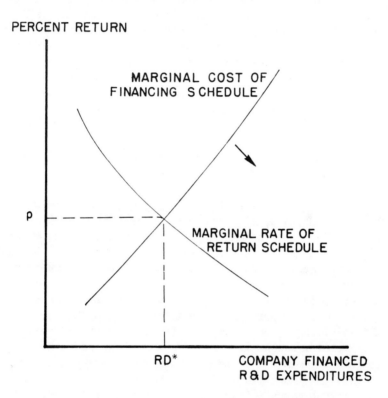

where I_T represents the firm's net private investments in R&D, and FRD represents the dollar value of the firm's federal R&D obligation. The marginal return to the firm's R&D investments is thus

$$d(\dot{A}/A)/d(I_T/Y) = \rho_1 + \rho_2 \, FRD. \tag{5.3}$$

If the firm receives no federal R&D obligations, then FRD = 0 and the marginal return is ρ_1 (it is ρ in Chapter 4). As FRD increases, the marginal rate of return to private R&D investments will increase or decrease according as ρ_2 is greater than or less than zero, ceteris paribus. It is hypothesized here that $\rho_2 < 0$.

A version of equation 5.3 was estimated, using ordinary least squares and data for the chemical, machinery, and transportation-equipment industries. The actual equation estimated included an inter-

TABLE 5.3

Estimated Rates of Return: Sample
of 33 Chemical Firms

Variable	Estimated Coefficient
Intercept	$\hat{\lambda}_1 = -0.300$* (-3.95)
D·intercept	$\hat{\lambda}_2 = -2.043$* (-9.44)
(I_T/Y)	$\hat{\rho}_1 = 0.031$† (2.18)
FRD·(I_T/Y)	$\hat{\rho}_2 = -0.006$ (-1.26)
D·(I_T/Y)	$\hat{\rho}_3 = 0.358$* (6.20)
D·FRD·(I_T/Y)	$\hat{\rho}_4 = -0.048$† (-2.60)
R^2	0.859
F-level	30.472

*Significant at the .01 level.
†Significant at the .05 level.
Note: t-statistics are reported in parentheses below each estimated coefficient.

cept and a slope term to account for the effect of firm size on the earned rate of return. That is, the estimating version of equation 5.3 is

$$\dot{A}/A = \lambda_1 + \lambda_2 \cdot D + \rho_1 (I_T/Y) + \rho_2 \text{FRD}(I_T/Y)$$
$$+ \rho_3 D \cdot (I_T/Y) + \rho_4 D \cdot \text{FRD}(I_T/Y) + \epsilon. \qquad (5.4)$$

For the smaller firms (as defined in Chapter 4), $D = 0$, and the estimated rate of return, without any federal R&D, is ρ_1, and with federal R&D, the rate of return is $(\rho_1 + \rho_2 \text{FRD})$. In the larger firms, $D = 1$, and the estimated rate of return, without any federal R&D, is $(\rho_1 + \rho_3)$, and with federal R&D, it is $[\rho_1 + \rho_3 + (\rho_2 + \rho_4)\text{FRD}]$. If $\rho_1 > 0$ and $\rho_3 > 0$, statistical evidence supporting the hypothesized

TABLE 5.4

Estimated Rates of Return: Sample of 34 Machinery Firms

Variable	Estimated Coefficient
Intercept	$\hat{\lambda}_1 = -0.094$ (-0.38)
D·intercept	$\hat{\lambda}_2 = -3.053^*$ (-4.33)
(I_T/Y)	$\hat{\rho}_1 = 0.001$ (0.02)
FRD·(I_T/Y)	$\hat{\rho}_2 = -0.005$ (-0.29)
D·(I_T/Y)	$\hat{\rho}_3 = 0.511^*$ (3.12)
D·FRD·(I_T/Y)	$\hat{\rho}_4 = -0.049\dagger$ (-2.22)
R^2	0.518
F-level	5.371

*Significant at the .01 level.
†Significant at the .05 level.

TABLE 5.5

Estimated Rates of Return: Sample of
19 Transportation-Equipment Firms

Variable	Estimated Coefficient
Intercept	$\hat{\lambda}_1 = -0.231$ (-0.24)
D·intercept	$\hat{\lambda}_2 = -7.610*$ (-2.88)
(I_T/Y)	$\hat{\rho}_1 = 0.104$ (0.25)
FRD·(I_T/Y)	$\hat{\rho}_2 = -0.001$ (-0.46)
D·(I_T/Y)	$\hat{\rho}_3 = 1.260\dagger$ (1.99)
D·FRD·(I_T/Y)	$\hat{\rho}_4 = -0.153$ (-1.20)
R^2	0.494
F-level	2.538

*Significant at the .05 level.
†Significant at the .10 level.

effect of federal R&D support in the smaller firms will be $\rho_2 < 0$, and in the larger firms, it will be $\rho_4 < 0$.

The estimated results for the sample of chemical firms are shown in Table 5.3. The return to private R&D in the smaller nonfederally supported firms is 3.1 percent (significant at the .05 level). The estimated value of $\hat{\rho}_2$ is not significantly different from zero, which suggests that the return earned in the smaller, federally supported firms is not different than that earned in the smaller nonfederally supported firms. The return to private R&D in the larger, nonfederally supported firms is 38.9 percent (significant at the .01 level). The return to the larger, federally supported firms is 38.0 percent, when FRD is evaluated at its mean.

These results do support the hypothesis that federal R&D does increase the efficiency of private R&D. The algebraic sign and the level of significance of $\hat{\rho}_4$ are constant with an expansion of the firm's own R&D (downward along its marginal rate-of-return function). The results also suggest that the magnitude of this effect is small, reducing the rate of return in the larger chemical firms by about 2 percent.

The estimated results for the sample of machinery firms are shown in Table 5.4. The estimated return to R&D in the smaller firms is zero, as was found in Chapter 4, regardless of the level of federal R&D support. The estimated return in the larger, nonfederally supported firms is 51.2 percent (significant at the .01 level). In the larger, federally supported firms it is 16.8 percent (when FRD is evaluated at its mean). In the machinery industry, the numerical impact of federal R&D is considerably greater than in the chemical industry. The marginal rate of return in the machinery industry has been reduced by 67 percent.

The estimated results for the transportation-equipment industry are reported in Table 5.5. As was estimated in Chapter 4, the rate of return in the smaller firms is zero. In the larger firms the return is 115.6 percent, regardless of the level of federal R&D support. The estimated coefficients on the FRD interaction terms are not significantly different from zero.

CONCLUSIONS

The analyses presented in this chapter question the efficiency with which technical knowledge created through the contracting process is transferred for societal consumption. The government admits that its bureaucratic establishments designed to meet the task of technological transfer resulting from contracted R&D projects are wanting. The empirical evidence presented suggests that performing firms are gaining few economies of production. The model posited to measure the impact of federal R&D on the firm's return to its own R&D is rather simple; nonetheless, the results imply that the impact of federal R&D on the productivity of private R&D is industry specific, and is relatively greater in larger firms. These findings may imply that the government must study the innovative process, on an industry and on a firm basis, before allocations are made. If not, government intervention may only create more inefficiencies.

The empirical evidence presented in this chapter, in contrast with previous studies, is consistent with the hypothesis that firms are capturing some benefits from conducting federal R&D. This conclusion is based on the finding that in the larger chemical and machinery firms, the rate of return earned on R&D decreases as federal R&D increases, implying that the firms may be expanding their own R&D programs. Since most firms are believed to underinvest in R&D, their expansion should have positive social benefits. It is difficult to evaluate this implication without having knowledge of the shape of each firm's or industry's marginal rate-of-return schedule; however, potential social gains are possible through federal R&D allocations. More study is clearly needed in this area.

There is an additional question to consider: Are there more efficient ways to stimulate private R&D (to lower the cost of financing) than through federal allocation programs? Presumably there are, if for no other reason than efficiency in the distribution of federal R&D explicitly assumes the government can correctly perceive the allocation scheme that will generate the greatest social benefits. Alternative programs must be considered and should be on the research agenda for developing an understanding of government participation in the innovation process.

6 SUMMARY

The analyses presented in this book represent a first attempt to investigate R&D activity in the manufacturing sector at a disaggregated level. It is argued in Chapter 1 that R&D is not a homogeneous activity; that R&D is an enormously heterogeneous activity, and must be studied at a disaggregated level if economists are to correctly develop public policy related to innovation and technological growth. R&D expenditures are disaggregated here in two dimensions: first, by character of use (basic, applied, and development), and second, by source of funding (company financed and federally financed).

This study is not without limitations or qualifications. The analytical analyses are based upon survey data that, by their very nature, are not pure. They correspond to a limited cross section of manufacturing firms for one fiscal year. No adjustments for lags are made. The data are open to criticism; however, no alternative source exists. If a disaggregated investigation is to be attempted, then certain data limitations may be inevitable. As well, the empirical models and analytic techniques are highly simplified. Consequently, the findings of this study are tentative and the conclusions are only suggestive.

The key variable examined in the empirical models is firm size. Size per se is an important focal variable to the extent that it proxies the influence of physical and financial economies of scale. Specifically, the role of size is considered as a determinant of the firm's decision to invest in alternative R&D activities; of the rate of return earned on these investments; and of the effectiveness of federal contracts and grants, in regard to firm-specific R&D activities.

The empirical findings in Chapters 3, 4, and 5 lead to four tentative conclusions. First, the importance of firm size as a determinant of intense R&D activity is limited. From the industries studied, the influence of size is relatively greater for basic research and least for development. Second, firm size is a prerequisite for successful R&D activity. Larger firms earn, at the margin, a relatively higher rate of return on their R&D expenditures than do smaller firms. This finding is true for total R&D as well as for basic, applied, and development expenditures. Third, the marginal return earned on basic-research expenditures is greater than on applied-research expenditures, and the marginal return on applied research is greater than on development expenditures. These findings support the proposition that the uncertainty associated with basic research is relatively the greatest. Fourth, larger firms appear to capture some of the social benefits from federally financed R&D. Specifically, the results in Chapter 5 support the hypothesis that in the larger firms, technical efficiencies generated through federal R&D contracts can be internalized by the performing firm, thus lowering its effective R&D costs and inducing expansion of its own R&D activity.

Additional research into other aspects of R&D is clearly warranted, especially at a disaggregated level. The conclusions presented in this chapter are, as noted above, subject to qualifications. Nevertheless, the conclusions suggest that public policy (especially that affecting firm size) may be misdirected if formulated on the view that R&D, and hence the overall innovation process, is a homogeneous activity.

A number of related areas deserve additional work. As indicated in Table 2.6, there is cross-industry variation in the allocation of R&D among alternative uses. Does this industry-allocation pattern reflect the technological environment of each industry? At a more disaggregated level, how can the interfirm differences in the allocation of R&D among basic, applied, and development be explained? As discussed in Chapter 4, evidence suggests that the productivity of R&D has collapsed in recent years, especially in smaller firms. Why? Can this phenomenon be attributed to a change in the firm's investment objectives? Has the nature of the innovation process changed over the past few decades? The analysis of Chapter 5 implicitly assumes that federal R&D dollars were in fact spent on federally related R&D projects. To what extent does federal R&D (especially as regards basic research) substitute for company R&D—that is, does crowding out take place? These important questions, and others, will certainly be on the agenda of researchers in the near future.

ём # APPENDIX A
SURVEY QUESTIONS

APPENDIX A

1. Approximately what percent of your total average annual research and development budget is company funded as opposed to federally funded? _____%

2. Approximately what percent of those research and development expenditures used by your company and funded by your company are directed toward:
   a. basic research _____%
   b. applied research _____%
   c. development _____%

3. Approximately what percent of those research and development expenditures used by your company and funded from federal sources are directed toward:
   a. basic research _____%
   b. applied research _____%
   c. development _____%

4. Approximately what average time lag exists between an initial research and development expenditure and the resulting improved products or improved processes? _____years

APPENDIX B
LISTING OF FIRMS IN THE SAMPLE
BY 4-DIGIT SIC CODES

APPENDIX B

LISTING OF FIRMS IN THE SAMPLE, BY FOUR-DIGIT SIC CODES

Four-Digit SIC Code	Firm Name
2000	General Foods Corp. General Mills Inc. Gerber Products Co. Quaker Oats Co. Standard Brands Inc. Universal Foods Corp.
2010	Esmark Inc. Valmac Inds. Inc.
2020	Kraft Inc.
2030	Green Giant Co. H. J. Heinz Co.
2046	A. E. Staley Mfg. Co.
2063	Holly Sugar Corp.
2065	Hershey Foods Corp.
2082	Adolph Coors Co.
2086	Coca-Cola Co.

100 / APPENDIX B

2111	U.S. Tobacco Co.
2200	Albany Intl. Corp. Burlington Inds. Inc. Dan River Inc. Lehigh Valley Inds. M. Lowenstein & Sons Inc. Reeves Brothers Inc. J. P. Stevens & Co. West Point-Pepperell
2270	Armstrong Cork Co. C. H. Masland & Sons
2400	Masonite Corp. Weyerhaeuser Co.
2600	Olinkraft Inc. Potlatch Corp. Scott Paper Co.
2649	Avery Intl. Bemis Co.
2711	Washington Post
2800	Allied Chemical Corp. Dow Chemical Ethyl Corp. W. R. Grace & Co. Koppers Co. Rohm & Haas Co. Union Carbide Corp.
2810	Air Products & Chemicals Inc. Airco Inc. Betz Laboratories Inc. Stauffer Chemical Co.
2820	Akzona
2830	Merck & Co. A. H. Robins Co. Schering-Plough Shaklee Corp. Sterling Drug Inc.

APPENDIX B / 101

2841	Colgate-Palmolive Co. Diversey Corp. Procter & Gamble Co.
2844	Gillette Co. International Flavors & Fragrances Revlon Inc.
2850	Cook Paint & Varnish De Soto Inc. Sherwin-Williams Co.
2860	Crompton & Knowles Corp. Nalco Chemical Co.
2890	Cabot Corp. Dexter Corp. Emery Inds. Inc. Ferro Corp. Sun Chemical Corp.
2911	Ashland Oil Inc. Atlantic Richfield Co. Cities Service Co. Continental Oil Co. Gulf Oil Corp. Shell Oil Co. Standard Oil Co. (Calif.) Standard Oil Co. (Indiana) Sun Oil Co. Union Oil Co. of California
2950	Flintkote Co.
3000	Armstrong Rubber Bandag Inc. Dayco Corp. B. F. Goodrich Co. Rubbermaid Inc. Uniroyal
3079	Monogram Inds. Inc.
3210	Corning Glass Works

3221	American Can Co. Continental Group
3241	Ideal Basic Inds. Inc.
3270	Ameron Inc.
3290	Norton Co. Vulcan Materials Co.
3310	Bethlehem Steel Corp. U.S. Steel
3330	Reynolds Metals Co.
3350	Belden Co.
3390	Kennametal Inc.
3449	Overhead Door Corp.
3494	Tyler Corp.
3510	Combustion Engineering Inc. Foster Wheeler Corp.
3520	Allis-Chalmers Corp. Deere & Co. Hesston Corp.
3531	Clark Equipment Co. FMC Corp. Koehring Co.
3533	Baker Intl. Corp. Dresser Inds. Inc. Hughes Tools Co.
3540	Wean United Inc.
3550	Black & Decker Mfg. Co. Emhart Corp. Ex-Cell-O Corp.

3560	Ansul Co. Barnes Group Inc. Briggs & Stratton Chicago Pneumatic Tool Co. Ingersoll-Rand Co. Parker Hannifin Corp. Riley Co. Twin Disc Inc.
3568	Combustion Equipment Assocs.
3570	California Computer Products Control Data Corp. A. B. Dick Co. Nashua Corp. Xerox Corp.
3573	Honeywell Inc. Sperry Rand Corp.
3580	Carrier Corp. Tecumseh Products Co. Trane Co.
3600	General Electric Co.
3610	Eltra Corp. Gould Inc.
3622	Leeds & Northrup Co.
3630	Singer Co.
3640	Thomas Inds. Inc.
3651	Zenith Radio Corp.
3662	Tracor Inc.
3670	Intel Corp.
3679	Ampex Corp. CTS Corp.

APPENDIX B

3699	Echlin Mfg. Co.
3711	General Motors Corp.
3713	Cummins Engine International Harvester Co.
3714	Bendix Corp. Champion Spark Plug Eaton Corp. Hayes-Albion Corp. Maremont Corp. TRW Inc.
3720	Lockheed Aircraft Corp.
3721	Beech Aircraft Corp. Boeing Co. McDonnell Douglas Corp.
3728	Rohr Inds.
3730	General Dynamics Corp.
3740	ACF Inds.
3760	Rockwell Intl. Corp.
3792	Coachmen Inds. Inc. Winnebago Inds.
3811	Beckman Instruments Inc. Technicon Corp.
3823	Fischer & Porter Co. Foxboro Co. Hewlett-Packard Co.
3841	American Sterilizer Co.
3843	Dentsply Intl. Inc.
3861	Minnesota Mining & Mfg. Co.
3870	Bulova Watch Co.

APPENDIX C
LOGARITHMIC REGRESSION RESULTS

TABLE C.1

Regression Results from Logarithmic Specifications
of Equations 3.1-3.4: Sample of 174 Firms

Variable	Total R&D	Basic	Applied	Development
Intercept	-3,790.529*	-173.554*	-203.674†	-3,422.796*
	(-7.59)	(-4.13)	(-2.24)	(-8.18)
Log S	1,762.825*	80.978*	99.105*	1,587.076*
	(8.22)	(4.48)	(2.54)	(8.83)
$(\text{Log S})^2$	-268.612*	-12.401*	-16.135*	-240.711*
	(-8.97)	(-4.92)	(5.45)	(-9.59)
$(\text{Log S})^3$	13.487*	0.627	0.895*	11.995*
	(9.88)	(5.45)	(3.60)	(10.48)
R^2	0.660	0.398	0.580	0.646
F-level	111.104	37.513	78.248	103.283

*Significant at .01 level.
†Significant at .05 level.
Note: t-statistics are reported in parentheses below each estimated coefficient.

TABLE C.2

Regression Results from Logarithmic Specifications of Equations 3.1-3.4 with 11 Separate Industry Effects: Sample of 174 Firms

Variable	Total R&D	Basic	Applied	Development
D20	120.74	3.30	-70.78	273.61
	(0.08)	(0.01)	(-0.14)	(0.27)
D20(log S)	-52.31	-1.05	36.61	-128.30
	(-0.08)	(-0.01)	(0.16)	(-0.28)
D20(log S)^2	6.92	0.07	-6.32	19.46
	(-0.07)	(0.01)	(-0.18)	(0.29)
D20(log S)^3	-0.24	0.003	0.37	-0.94
	(-0.53)	(0.01)	(0.21)	(-0.28)
D21, 22	642.05	143.79	153.98	187.65
	(0.27)	(0.28)	(0.18)	(0.11)
D21, 22(log S)	-323.78	-72.49	-78.43	-93.67
	(-0.28)	(-0.29)	(-0.18)	(-0.11)
D21, 22(log S)^2	52.51	11.99	13.11	15.29
	(0.28)	(0.29)	(0.19)	(0.11)
D21, 22(log S)^3	-2.88	-0.65	-0.72	-0.81
	(-0.29)	(-0.30)	(-0.19)	(-0.11)
D24, 26, 27	4,452.16	110.80	-1,193.62	5,534.99
	(0.42)	(0.05)	(-0.29)	(0.71)
D24, 26, 27 (log S)	-1,974.14	-46.53	56.53	-2,491.16
	(-0.42)	(-0.04)	(0.32)	(-0.73)
D24, 26, 27 (log S)^2	288.57	6.49	-88.61	370.69
	(0.41)	(0.04)	(-0.34)	(0.74)
D24, 26, 27 (log S)^3	-13.86	-0.30	4.65	-18.21
	(-0.41)	(-0.04)	(0.37)	(-0.74)
D28	-1,038.28	-133.39	-234.24	-675.57
	(-1.59)	(-0.93)	(-0.97)	(-1.44)

continued

TABLE C.2 (continued)

Variable	Total R&D	Basic	Applied	Development
D28 (log S)	521.241**	69.30	115.48	339.31
	(1.73)	(1.04)	(1.03)	(1.55)
D28 (log S)^2	-86.88†	-11.96	-19.02	-56.41**
	(-1.89)	(-1.19)	(-1.12)	(-1.70)
D28 (log S)^3	4.88†	0.69	1.07	3.14**
	(2.15)	(1.38)	(1.28)	(1.92)
D29	3,173.78	-653.23	2020.77	1,827.82
	(0.67)	(-0.63)	(1.15)	(0.54)
D29 (log S)	-1,165.79	252.33	-752.64	-673.57
	(-0.65)	(0.64)	(-1.13)	(-0.52)
D29 (log S)^2	137.87	-31.98	90.79	80.04
	(0.62)	(-0.65)	(1.09)	(0.50)
D29 (log S)^3	-5.22	1.33	-3.54	-3.05
	(-0.57)	(0.67)	(-1.05)	(-0.46)
D30, 32	3,263.09	328.07	585.64	2,334.48
	(1.35)	(0.62)	(0.65)	(1.34)
D30, 32 (log S)	-1,552.99	-156.28	-279.22	-1,111.14
	(-1.38)	(-0.63)	(-0.67)	(-1.36)
D30, 32 (log S)^2	241.62	24.39	43.41	172.91
	(1.39)	(0.64)	(0.67)	(1.38)
D30, 32 (log S)^3	-12.22	-1.24	-2.18	-8.75
	(-1.38)	(-0.64)	(-0.66)	(-1.37)
D33, 34	215.60	70.08	-18.29	-124.44
	(0.17)	(0.25)	(-0.04)	(-0.13)
D33, 34 (log S)	-90.55	-35.43	13.30	65.14
	(-0.16)	(-0.28)	(0.06)	(0.16)

continued

TABLE C.2 (continued)

Variable	Total R&D	Basic	Applied	Development
D33, 34$(\log S)^2$	11.48 (0.14)	5.74 (0.31)	-2.99 (-0.09)	-11.24 (-0.18)
D33, 34$(\log S)^3$	-0.38 (-0.09)	-0.29 (-0.32)	0.22 (0.14)	0.65 (0.22)
D35	-2,750.41* (-3.56)	-104.34 (-0.62)	-514.42** (-1.81)	-2,131.66* (-3.86)
D35$(\log S)$	1,441.89* (3.94)	54.17 (0.67)	272.44† (1.99)	1,115.28* (4.21)
D35$(\log S)^2$	-249.26* (-4.32)	-9.25 (-0.73)	-47.63† (-2.22)	-192.38* (-4.61)
D35$(\log S)^3$	14.242* (4.77)	0.52 (0.79)	2.75† (2.48)	10.97* (5.08)
D36	-2,573.83† (-2.42)	-36.13 (-0.15)	-707.68** (-1.78)	-1,830.02† (-2.38)
D36$(\log S)$	1,248.40* (2.64)	17.99 (0.17)	350.44† (1.99)	879.97* (2.58)
D36$(\log S)^2$	-199.44* (-2.95)	-2.97 (-0.20)	-57.16† (-2.27)	-139.31* (-2.85)
D36$(\log S)^3$	10.58* (3.38)	0.16 (0.24)	3.08* (2.65)	7.33* (3.25)
D37	-4,308.20* (-5.93)	-240.10 (-1.50)	115.44 (0.43)	4,184.46* (-7.97)
D37$(\log S)$	2,009.48* (6.98)	112.70** (1.78)	-31.66 (-0.30)	1,928.78* (9.26)
D37$(\log S)^2$	-310.81* (-8.42)	-17.45† (-2.15)	0.86 (0.06)	-294.26* (-11.03)

continued

TABLE C.2 (continued)

Variable	Total R&D	Basic	Applied	Development
$D37(\log S)^3$	15.99*	0.89*	0.20	14.90*
	(10.45)	(2.65)	(0.35)	(13.47)
$D38$	2,394.99	-173.80	325.32	2,243.47
	(1.16)	(-0.38)	(0.42)	(1.50)
$D38(\log S)$	-1,188.85	90.25	-157.43	-1,121.67
	(-1.20)	(0.41)	(-0.43)	(-1.56)
$D38(\log S)^2$	189.60	-15.46	24.03	181.03
	(1.21)	(-0.45)	(0.41)	(1.60)
$D38(\log S)^3$	-9.50	0.87	-1.11	-9.27
	(-1.17)	(0.49)	(-0.37)	(-1.59)
R^2	0.973	0.676	0.862	0.979
F-level	109.886	6.320	18.844	142.617

*Significant at .01 level.
†Significant at .05 level.
**Significant at .10 level.

BIBLIOGRAPHY

Abramowitz, Moses. 1956. "Resource and Output Trends in the U.S. Since 1870." American Economic Review 46: 5-23.

Adams, William J. 1970. "Firm Size and Research Activity: France and the United States." Quarterly Journal of Economics 84: 386-409.

Agnew, Carson E., and Donald E. Wise. 1978. "The Impact of R & D on Productivity: A Preliminary Report." Paper presented at the Southern Economic Association meetings.

Arrow, Kenneth J. 1962. "Economic Welfare and the Allocation of Resources for Invention." In The Rate and Direction of Inventive Activity, pp. 609-25. Princeton: Princeton University Press, for the National Bureau of Economic Research.

Ault, Leonard A. 1979. "NASA Technology and Utilization Program." In Federal R & D and Scientific Innovation, edited by Leonard A. Ault and W. Novis Smith, pp. 65-78. Washington, D.C.: American Chemical Society.

Brown, R. L., J. Durbin, and J. M. Evans. 1975. "Techniques for Testing the Constancy of Regression Relationships Over Time." Journal of the Royal Statistical Society (series B) 37: 149-63.

Brumm, Harold J., Jr. and John Hemphill. 1976. "The Role of Government in the Allocation of Resources to Technological Innovation." Final report submitted to the Office of National R & D Assessment, National Science Foundation. Mimeographed.

Business Week. 1976. "The Breakdown of U.S. Innovation." February 16, pp. 56-68.

_____. 1978. "Vanishing Innovation." July 3, pp. 46-54.

Comanor, William S. 1967. "Market Structure, Product Differentiation, and Industrial Research." Quarterly Journal of Economics 81: 639-57.

Committee for Economic Development 1980. Stimulating Technological Progress. New York: Heffernan Press.

Danhoff, Clarence H. 1968. Government Contracting and Technological Change. Washington, D.C.: The Brookings Institution.

Demsetz, Harold. 1969. "Information and Efficiency: Another Viewpoint." Journal of Law and Economics 1: 1-22.

Denison, Edward F. 1979. "Explanations of Declining Productivity Growth." Survey of Current Business 59: 1-24.

Enos, John L. 1962. "Invention and Innovation in the Petroleum Refining Industry." In The Rate and Direction of Inventive Activity, pp. 299-322. Princeton: Princeton University Press, for the National Bureau of Economic Research.

Fisher, Franklin M., and Peter Temin. 1973. "Returns to Scale In Research and Development: What Does the Schumpeterian Hypothesis Imply?" Journal of Political Economy 81: 56-70.

Freeman, Christopher. 1974. The Economics of Industrial Innovation. London: Penguin Books.

Gold, Bela. 1971. Explorations in Managerial Economics: Productivity, Costs, Technology and Growth. London: Macmillan.

Goldberg, Lawrence. 1978. "Federal Policies Affecting Industrial Research and Development." Paper presented at the Southern Economic Association meetings.

Grabowski, Henry G., and Dennis C. Mueller. 1970. "Industrial Organization: The Role and Contribution of Econometrics." American Economic Review: Papers and Proceedings 70: 100-4.

Griliches, Zvi. 1958. "Research Costs and Social Returns: Hybrid Corn and Related Innovations." Journal of Political Economy 66: 419-31.

_____. 1964. "Research Expenditures, Education, and the Aggregate Agriculture Production Function." American Economic Review 54: 961-74.

_____. 1973. "Research Expenditures and Growth Accounting." In Science and Technology in Economic Growth, edited by B. R. Williams, pp. 59-83. New York: John Wiley & Sons.

_____. 1979. "R&D and the Productivity Slowdown." Paper presented at the American Economic Association meetings.

_____. 1980. "Returns to Research and Development Expenditures in the Private Sector." In New Developments in Productivity Measurement, edited by John W. Kendrick and Beatrice Vaccara, pp. 419-62. New York: National Bureau of Economic Research.

Hamberg, Daniel. 1964. "Size of Firm, Oligopoly, and Research." Canadian Journal of Economics and Political Science 30: 62-75.

_____. 1966. R&D: Essays on the Economics of Research and Development. New York: Random House.

Heller, Robert H., and Moshin S. Khan. 1979. "The Demand for Money and the Term Structure of Interest Rates." Journal of Political Economy 87: 109-29.

Hodgson, John S., and Alexander B. Holmes. 1977. "Structural Stability of International Capital Mobility: An Analysis of Short-Term U.S.-Canadian Bank Claims." Review of Economics and Statistics 59: 465-73.

Hollander, Samuel. 1965. The Sources of Increased Efficiency. Cambridge: MIT Press.

Horowitz, Ira. 1962. "Firm Size and Research Activity." Southern Economic Journal 28: 298-301.

Hu, Sheng Cheng. 1973. "On the Incentive to Invent: A Clarificatory Note." Journal of Law and Economics 16: 169-77.

Jewkes, John, David Sawers, and Richard Stillerman. 1969. The Sources of Invention. 2d ed. London: Macmillan.

Kay, Neil M. 1979. The Innovating Firm. New York: St. Martin's Press.

Kamien, Morton I., and Nancy L. Schwartz. 1975. "Market Structure and Innovation." Journal of Economic Literature 13: 1-37.

Kennedy, Charles, and A. P. Thirwall. 1972. "Surveys in Applied Economics." The Economic Journal 82: 11-72.

Khan, Moshin S. 1974. "The Stability of the Demand-for-Money Function in the United States 1901-1965." Journal of Political Economy 82: 1205-19.

Kirzner, Israel M. 1973. Competition and Entrepreneurship. Chicago: University of Chicago Press.

Knight, Frank M. 1921. Risk, Uncertainty and Profit. Boston: Houghton Mifflin.

Kochanowski, Paul S., and Henry R. Hertzfeld. 1980. "Often Overlooked Factors in Measuring the Rate of Return to Government R&D Expenditures." Policy Analysis 6.

Link, Albert N. 1977. "On the Efficiency of Federal R&D Spending: A Public Choice Approach." Public Choice 31: 129-33.

_____. 1978. "Rates of Induced Technology from Investments in Research and Development." Southern Economic Journal 45: 370-79.

_____. 1980. "Firm Size and Efficient Entrepreneurial Activity: A Reformulation of the Schumpeter Hypothesis." Journal of Political Economy 88: 771-82.

Link, Albert N., and Stephen O. Morrell. 1980. "Political Competition for Government Funds: Research and Development Obligations." Public Finance Quarterly 8: 57-67.

Loeb, Peter D., and Vincent Lin. 1977. "Research and Development in the Pharmaceutical Industry: A Specification Error Approach." Journal of Industrial Economics 26: 45-51.

Machlup, Fritz. 1962. "The Supply of Inventors and Inventions." In

The Rate and Direction of Inventive Activity, pp. 143-67. Princeton: Princeton University Press, for the National Bureau of Economic Research.

Mansfield, Edwin. 1964. "Industrial Research and Development Expenditures: Determinants, Prospects, and Relation to Size of Firm and Inventive Output." Journal of Political Economy 72: 319-40.

———. 1965. "Rates of Return from Industrial Research and Development." American Economic Review 55: 310-22.

———. 1968a. The Economics of Technological Change. New York: Norton.

———. 1968b. Industrial Research and Technological Innovation—An Econometric Analysis. New York: Norton.

———. 1969. "Industrial Research and Development: Characteristics, Costs and Diffusion of Results." American Economic Review 59: 65-71.

———. 1972. "Contributions of Research and Development to Economic Growth of the United States." In Research and Development and Economic Growth/Productivity, pp. 21-36. Washington, D.C.: Government Printing Office.

———. 1976. "Federal Support of R&D Activities in the Private Sector." In Priorities and Efficiency in Federal Research and Development, pp. 85-115. Washington, D.C.: Government Printing Office.

Mansfield, Edwin, John Rapoport, Anthony Romeo, Samuel Wagner, and George Beardsley. 1977. "Social and Private Rates of Return from Industrial Innovations." Quarterly Journal of Economics 91: 221-40.

Markham, Jesse W. 1962. "Inventive Activity: Government Controls and the Legal Environment." In The Rate and Direction of Inventive Activity, pp. 587-608. Princeton: Princeton University Press, for the National Bureau of Economic Research.

———. 1965. "Market Structure, Business Conduct, and Innovation." American Economic Review: Papers and Proceedings 55: 323-32.

Merrill, Stephen A. 1979. "The Political Nature of Civilian R&D Management." In Federal R&D and Scientific Innovation, edited by Leonard A. Ault and W. Novis Smith, pp. 3-13. Washington, D.C.: American Chemical Society.

Minasian, Jora R. 1962. "The Economics of Research and Development." In <u>The Rate and Direction of Inventive Activity</u>, pp. 93-142. Princeton: Princeton University Press, for the National Bureau of Economic Research.

———. 1969. "Research and Development, Production Functions, and Rates of Return." <u>American Economic Review</u> 59: 80-85.

Mueller, Dennis C. 1966. "Patents, Research and Development, and the Measurement of Inventive Activity." <u>Journal of Industrial Economics</u> 15: 26-37.

———. 1967. "The Firm Decision Process: An Econometric Investigation." <u>Quarterly Journal of Economics</u> 81: 58-87.

Nason, Howard K. 1979. "The Environment for Industrial Innovation in the United States." In <u>Technological Innovation: Government/Industry Cooperation</u>, edited by Arthur Gerstenfeld, pp. 69-79. New York: John Wiley & Sons.

Nason, Howard K., Joseph A. Steger, and George E. Manners. 1978. <u>Support of Basic Research by Industry</u>. Washington, D.C.: National Science Foundation.

National Science Foundation. 1978. <u>National Patterns of R&D Resources</u>. Washington, D.C.: Government Printing Office.

———. 1979a. <u>Federal Funds for Research and Development</u>. Washington, D.C.: Government Printing Office.

———. 1979b. <u>Research and Development in Industry,</u> 1977. Washington, D.C.: Government Printing Office.

Ng, Yew-Kwang. 1971. "Competition, Monopoly, and the Incentive to Invent." <u>Australian Economic Papers</u> 10: 52-65.

Peltzman, Sam. 1973. "An Evaluation of Consumer Protection Legislation: The 1962 Drug Amendments." <u>Journal of Political Economy</u> 81: 1049-86.

Phlips, Louis. 1971. "Research." In <u>Effects of Industrial Concentration: A Cross Section Analysis for the Common Market</u>, pp. 119-42. Amsterdam: North-Holland.

Pigou, A. C. 1932. The Economics of Welfare. 3rd ed. London: Macmillan.

Quandt, Richard E. 1960. "Tests of Hypothesis That a Linear Regression System Obeys Two Separate Regimes." Journal of the American Statistical Association 55: 324-30.

Quesenberry, William O. 1979. "Patents and Technology Transfer." In Federal R&D and Scientific Innovation, edited by Leonard A. Ault and W. Novis Smith, pp. 79-86. Washington, D.C.: American Chemical Society.

Scherer, Frederick, M. 1965a. "Firm Size, Market Structure, Opportunity, and the Output of Patented Inventions." American Economic Review 55: 1097-1125.

_____. 1965b. "Size of Firm, Oligopoly, and Research: A Comment." Canadian Journal of Economics 31: 256-66.

_____. 1973. "Research and Development Returns to Scale and The Schumpeterian Hypothesis: Comment." Reprint. International Institute of Management.

Schlie, Theodore W. 1979. "A Summary of the Background Process, and Results of President Carter's Domestic Policy Review on Industrial Innovation." Paper presented at the Southern Economic Association meetings.

Schmookler, Jacob. 1952. "The Changing Efficiency of the American Economy, 1869-1938." Review of Economics and Statistics 34: 214-31.

_____. 1959. "Bigness, Fewness, and Research." Journal of Political Economy 67: 628-32.

Schon, Donald A. 1967. Technology and Change: The New Heraclitus. Oxford, Pergamon.

Schultz, Theodore W. 1975. "The Value of the Ability to Deal with Disequilibria." Journal of Economic Literature 13: 827-46.

Schumpeter, Joseph A. 1939. Business Cycles. New York: McGraw-Hill.

_____. 1947. Capitalism, Socialism, and Democracy. New York: Harper & Brothers.

Sheils, Merrill, et al. 1979. "Innovation: Has America Lost Its Edge?" Newsweek, June 4, pp. 58-68.

Solow, Robert M. 1957. "Technical Change and the Aggregate Production Function." Review of Economics and Statistics 57: 312-20.

Stern, Robert M., Christopher F. Baum, and Mark N. Greene. 1979. "Evidence on Structural Change in the Demand for Aggregate U.S. Imports and Exports." Journal of Political Economy 87: 179-92.

Stroetmann, Karl A. 1979. "Innovation in Medium and Small Industrial Firms." In Technological Innovation: Government/Industry Cooperation, edited by Arthur Gerstenfeld, pp. 93-103. New York: John Wiley & Sons.

Terleckyj, Nestor E. 1974. Effects of R&D on the Productivity Growth of Industries: An Exploratory Study. Washington, D.C.: National Planning Association.

Usher, Abbott P. 1954. A History of Mechanical Inventions. Cambridge: Harvard University Press.

Villard, Henry H. 1958. "Competition, Oligopoly and Research." Journal of Political Economy 66: 483-97.

Wagner, Leonore, 1968. "Problems in Estimating Research and Development Investments and Stock." Proceedings of the American Statistical Association: Business and Economic Statistics Section, pp. 189-98.

Worley, James S. 1961. "Industrial Research and the New Competition." Journal of Political Economy 69: 183-86.

Young, Lewis M. 1979. "To Revive Research and Development." Business Week, September 17, p. 21.

INDEX

Abramowitz, M., 1
Adams, W., 30
Agnew, C., and D. Wise, 55, 59, 82
applied research: construct validity of, 12-13; definition of, 4, 49; federal funding and, 75; as a function of firm size, 31, 41-49; relation to productivity, 62; returns to, 49, 70-71, 91-92; total amount spent for, 7, 12, 18 (see also research and development)
Arrow, K., 25, 78
Ault, L., 81

basic research: construct validity of, 12; definition of, 4; federal funds and, 75; as a function of firm size, 31, 41-49; productivity, 62; returns to, 49, 70-71, 73, 92; total amount spent for, 7, 12, 18 (see also research and development)
Brumm, H., and J. Hemphill, 79
Bureau of Labor Statistics, 58
Business Week, 11

Carter, J., 81
Caterpillar Tractor, 48
character of use of R&D, 48 (see also applied research, basic research, data, development, research and development)
chemical industry: character of use of R&D in, 48; intensity of R&D in, 41-48; R&D as a function of a size in, 29, 30, 41-48; rates of return to R&D in, 62, 69-70, 87-88, 89; representatives of in sample, 17; threshold size level of, 69
Comanor, W., 28-32
Compustat, 11, 17, 21, 32, 58
confidence limits: Pyke's modified Kolmogorov-Smirnov statistic, 68

data: available on R&D, 3-4; description of, 11, 17-23; coverage ratios, 17; industry distribution of, 17, 41; sample of respondents, 13-17; survey, 12
Deere, 48
Demsetz, H., 25
Department of Agriculture, 75 (see also government)
Department of Defense, 17, 75-77;

Navy, 75 (see also government)
Department of Energy, 77 (see also government)
Department of Education, 77 (see also government)
development: construct validity of, 13; definition of, 4, 49; federal funding and, 75; as a function of firm size, 31, 41-49; relationship to productivity, 62; return to, 49, 70-71, 73, 92; total amount spent for, 7, 12, 18 (see also research and development)
Dow Chemical, 37
drug industry: R&D as function of size in, 29

economies of scale: in production, 30, 88; in R&D, 3, 25-26, 72, 80
elasticity: of R&D in regard to firm size, 28; output, 52-53
electrical-equipment industry: representatives of in sample, 17; R&D financing of in, 17; R&D as function of size in, 29
Enos, J., 21
entrepreneur: role of, 2-3, 64-65, 79, 81
expenditures on R&D: 51-52; previous studies, 52-55 (see research and development: total expenditures)

Fabricant, S., 1
Federal Food, Drug, and Cosmetic Act (1938), 48
Fisher, F., and P. Temin, 30-31
food and tobacco industry: R&D as function of size in, 29
Ford Motor Co., 13
Freeman, C., 3

General Electric, 32
General Motors, 13, 32, 49
glass industry: R&D as function of size in, 30
Gold, B., 57
Goldberg, L., 82
government (federal): benefits of funding R&D, 80-82; and efficiency of R&D, 88, 89; R&D financing by, 6-7, 17, 21, 75-89, 92; role in innovation by R&D, 75-80
Grabowski, H., and D. Mueller, 30
Griliches, Z., 54-55, 59, 82

Hamberg, D., 28, 32
Hollander, S., 51
Horowitz, I., 28
Hu, Sheng Cheng, 25

industrial R&D (see research and development)
industrial trends in R&D, 7 (see also research and development)
information (see technical knowledge)
innovation: characteristics of, 3; definition of, 2-3; determinants of, 2-3, 25-26; Domestic Policy Review of, 81-82; effect of on technological change, 2; intensity of, 32-49; measurements of, 26, 31, 32; process (type of), 53; process of, 21, 51, 57, 78-79, 92; product (type of), 53; and size, previous studies, 27-31; Technology Utilization Program on, 81
intensity: of innovation, 32-49; of R&D, 53
International Business Machines, 13, 48
International Harvester, 48
invention: cause of, 3; impact on technological change, 2, 51; lag of, 21; sources of, 2-3
Investor's Management Sciences, Inc., 11

122

Jewkes, J., D. Sawers, and R. Stillerman, 2-3

Kamien, M., and N. Schwartz, 80
Kirzner, I., 64
Knight, F., 3
Kochanowski, P., and Hertzfeld, H., 83

labor: share in total output, 58
lags: in R&D, 21-23, 91
Link, A., 54, 68-69, 80
Link, A., and S. Morrell, 80
Loeb, P., and V. Lin, 30

Machlup, F., 57
machinery industry: character of use of R&D, 48; innovative intensity of, 48; R&D as a function of size in, 29, 48; rates of return to R&D, 64, 70-71, 88; representatives of in sample, 17; threshold size of, 69
Mansfield, E., 1, 29, 48, 53-54
manufacturing sector: leading industries of, 17; representatives of in sample, 13-17; studies of, 53, 91
marginal rate of return (see rates of return)
Markham, J., 30
Merrill, S., 81
Minasian, J., 52-53
monopoly: demand for innovation, 25
Mueller, D., 29

Nason, H., 2
Nason, H., J. Steger, and G. Manners, 12
National Aeronautics and Space Administration (NASA), 17, 75; Technology Utilization Program, 81

National Science Foundation, 3-4, 11, 12, 18, 21, 27
National Science Foundation Act (1947), 75
national trends in R&D, 6-7 (see also research and development)
Ng, Yew-Kwang, 25

Pareto efficiency, 78
Patent Abstract Bulletin, 81
patents, 27
petroleum-products industry: lag between invention and innovation in, 21; R&D as function of size in, 29; representatives of in sample, 17
pharmaceutical industry: innovative behavior in, 48; R&D as function of size in, 30
Philips, L., 30
Pigou, A. C., 52, 79-80
policy for R&D, 26, 29, 49, 53, 73, 89, 92
President's Domestic Policy Review on Industrial Innovation, 81
price: deflator, 58; indices, 58
primary-metal industry: R&D as function of size in, 29
process innovations, 54, 83
Procter and Gamble, 37
product innovations, 54, 83
production frontiers, 55
production function: of the firm, 52, 53, 54, 83; Cobb-Douglas, 53, 55
productivity: determinants of, 4; growth in, 1, 51-55, 57, 92; of R&D, 55, 83, 88-89; relationship of to R&D, 52-53, 54, 55, 62; residual measures of, 51, 83
public goods, 78

Quesenberry, W., 80-81

R&D (see research and development)
rates of return: to R&D, 52-55;

to scale, 29; and size, 69-72; social versus private, 53, 54, 78, 79, 92; statistical estimation of, 55-58

regulation of firms, 55

research and development: allocation of resources to, 29, 75, 80; applied research, 4, 7, 12-13, 18, 31, 41-49, 62, 70-71, 75, 92; basic research, 4, 7, 12-13, 18, 31, 41-49, 62, 70-71, 73, 92; character of use, 4, 7, 18, 75, 91, 92; definition of, 4; determinants of, 27-31; development, 4, 7, 12-13, 18, 31, 41-49, 62, 70-71, 73, 92; efficiency of, 88; elasticity of, 28; expenditures by industry, 18; industrial trends in, 7; inputs, 27; lags associated with, 21-23, 91; national trends, 4-7; productivity of, 55, 83, 88; rates of return on, 52-55, 64, 65, 82, 91-92; relationship of to productivity, 52-53, 54, 55, 62; sources of funding of, 4-7, 21, 75, 91; statistics on, 11 (see also applied research, basic research, development, lags, productivity, rates of return)

risk: determinants of, 3; in R&D, 48, 79

Scherer, F., 29, 32, 48
Schmookler, J., 1, 27
Schon, D., 57
Schultz, T., 65
Schumpeter, J., 2, 25, 30, 49, 57, 65
Schumpeterian hypothesis, 25-26, 30-31, 65
self-interest hypothesis of R&D allocation, 80
size of firm: distribution of sample, 13; effect of on R&D, 3, 25-26, 49, 92; and innovative intensity, 41-49; R&D and previous studies, 27-31

social gain from R&D, 80-88 (see also rates of return)

Solow, R., 1, 51, 52
spillover effects: of R&D, 53, 80; of technical knowledge, 53
Standard and Poor's, 11
steel industry: R&D as function of size in, 30
Stroetmann, K., 2
structural change: Brown-Durbin-Evans test for, 68; Quandt's log-likelihood ratio test, 68-69
survey: description of, 11-12

technical capital, 53, 55
technical knowledge: origin of, 51, 80; spillover of, 53, 78; stock of, 54, 78, 81; transfer of, 81, 88
technological change: definition of, 1; determinants of, 1, 2, 7, 49, 51; history of, 2-3
technology: origin of, 51
Technology Utilization Program, 81
Terleckyj, N., 53, 82-83
transportation-equipment industry: character of use of R&D, 48; innovative intensity of, 48; R&D financing of, 17; R&D as function of size in, 49; rates of return to R&D, 64, 71-72, 88; representatives of in sample, 13-17; threshold size of, 69

uncertainty: determinant of, 3; in R&D, 79; types of, 3
Union Carbide, 37
United States: technological superiority of, 7
United States Congress, 75, 81
Usher, A., 2

Villard, H., 27
Worley, J., 28
Xerox, 32

ABOUT THE AUTHOR

ALBERT N. LINK is assistant professor of Economics at Auburn University. Dr. Link has published widely in the area of economics. His articles have appeared in such academic journals as the <u>Journal of Political Economy</u>, <u>Public Choice</u>, <u>Public Finance Quarterly</u>, and the <u>Southern Economic Journal</u>. He is also the author of several research bulletins prepared for the Department of the Interior.

Dr. Link holds a B.S. (1971) in mathematics from the University of Richmond, and a Ph.D. (1976) in economics from Tulane University.